Tendons

Edited by Hasan Sözen

Published in London, United Kingdom

IntechOpen

Supporting open minds since 2005

Tendons
http://dx.doi.org/10.5772/intechopen.73878
Edited by Hasan Sözen

Contributors
Rita de Cassia Marqueti, Fabricio Reichert Barin, Ivo Vieira de Sousa Neto, Gracielle Vieira Ramos, Kwang-Il Lee, Kwang-Won Lee, Ju-Woong Jang, Janina Burk, Nazım Karahan, Barış Yılmaz, Murat Kaya, Justin Yousef, Tawheed Ahmad, Summaira Jan, Saima Rashid, Mayur Nayak, Rahul Yadav, Stephanie C. Torres-Ayala, Jorge A. Vidal, Amanda P. Marrero-González, Alvaro Bravo-Martínez, Luis R. Rodriguez-Ortíz, Hasan Sözen

First published in London, United Kingdom, 2019 by IntechOpen
IntechOpen is the global imprint of INTECHOPEN LIMITED, registered in England and Wales, registration number: 11086078, 7th floor, 10 Lower Thames Street, London, EC3R 6AF, United Kingdom
Printed in Croatia

British Library Cataloguing-in-Publication Data
A catalogue record for this book is available from the British Library

Additional hard and PDF copies can be obtained from orders@intechopen.com

Tendons
Edited by Hasan Sözen
p. cm.
Print ISBN 978-1-83962-985-3
Online ISBN 978-1-83962-986-0
eBook (PDF) ISBN 978-1-83962-987-7

For EU product safety concerns:
IN TECH d.o.o., Prolaz Marije Krucifikse Kozulić 3, 51000 Rijeka, Croatia, info@intechopen.com or visit our website at intechopen.com.

Meet the editor

Dr. Hasan Sözen is an Assistant Professor in the School of Physical Education and Sport at Ordu University (Ordu, Turkey). His primary research interest includes sport and exercise physiology. Dr. Sözen received his PhD from the Health Science Institute, Department of Physical Education and Sport at Ondokuz Mayıs University (Samsun, Turkey). He completed his post-doctoral fellowship at the Department of Biomedical Sciences for Health at the University of Milan (Milan, Italy), and his thesis was titled "Electromechanical delay components assessment to disclose age, training status and gender effects on skeletal muscle electromechanical behavior during contraction: new insights from an electromyographic, mechanomyographic and force combined approach". The fellowship was supported by the Scientific and Technological Research Council of Turkey (TUBİTAK).

Contents

Preface

The muscles and skeletal system are vital for performing everyday functions. Tendons are round, oval, or flat tissue extending between muscle and bone. Tendons are differentiated muscles that connect to bones. They provide joint movements resulting from contraction and relaxation of muscles. Tendons are highly resistant to tensile force but have a flexible structure. They also have some extension properties. In this manner, they transmit the tension generated by the muscle to the bones and tendons adapt perfectly to the joint regions and bone circumference. Some tendons contain bony or cartilaginous sesamoid bones. These bones allow the tendons to adapt to the bone surface. The tensile strength of the tendons is similar to the bone, and a 1 cm thick tendon can withstand a load of 600–1000 kg.

The musculoskeletal system can be likened to the columns that carry the body. A weak system will cause you to experience discomfort in different parts of your body after a while. A muscle and skeletal system that is strong enough to perform vital functions is extremely important. Walking, running, performing sports, and even moving comfortably, depends on it. With the increasing importance placed on human health, the average life expectancy has started to increase. Improving the quality of life for this longer lifespan has become very important.

The book contains an introductory chapter, which is followed by an overview of the structure and classification of tendons by Kaya et al., including an explanation of their complex structure. The third chapter is about the imaging of tendons by Torres-Ayala et al., where we learn that magnetic resonance imaging and ultrasound are useful radiological methods that allow adequate evaluation of tendon anatomy and integrity. The fourth chapter is on exercise and a tendon remodeling mechanism by de Cassia Marqueti et al. The fifth chapter describes patellar tendinopathy, which is a source of anterior knee pain, characterized by pain localized to the inferior pole of the patella. The author describes patellar tendinopathy as jumper's knee. The sixth chapter by Burk summarizes a very important, yet mostly underestimated subject: mechanisms of action of multipotent mesenchymal stromal cells in tendon disease. The chapter provides an understanding of successful treatment approaches to fully exploit the regenerative potential of the multipotent mesenchymal stromal cells. The seventh chapter is by Yousef and it introduces the physiology of flexor tendon healing and rationale for treatment protocols. This is an interesting chapter on the management of flexor tendon injuries of the hand by Ahmad et al. We use our hands for carrying out most of our daily activities, but these activities can make our hands vulnerable for trauma. The last chapter is written by Lee et al. and it describes an interesting study performed with the injectable rhBMP-2-containing collagen gel for tendon healing in a rabbit extra-articular bone tunnel model. This book may contain errors despite our obsessive reviews and efforts. But all in all, I think that it provides the reader with interesting up-to-date data while summarizing information about tendons.

I want to thank all the authors of this book for their amazing work and our Author Service Manager Ms Rozmari Marijan, without whom I would not have been able to edit this book.

I hope that this book will be useful for anyone who wants to read about new perspectives on tendons. I also hope that it will inspire researchers working in this field.

Dr. Hasan Sözen
University of Ordu,
Physical Education and Sport,
Ordu, Turkey

Chapter 1

Introductory Chapter: Tendons

Hasan Sözen

1. Introduction

The tendons act as a mechanical bridge. Tendons allowing muscle strength to pass to the bones and joints, it also allows the muscle to contract and target movement. There are different types of tendons that reflect muscle morphology and specific functions. Tendon tissue includes all muscle tissue, not just the terminal or starting area of each muscle. The binding layers (epimysium, perimysium, and endomysium) combine in a single organization to contact one or more fixed bone points. There is a contraction fiber in the same tendon near the muscle. It affects the muscle-tendon, and thus the tendon affects the functional function of the muscle. In the context of manual therapy, rehabilitation or surgery, it is important to consider these close relationships between anatomy and function. Tendon tissue can adapt its cellular structure to pathological or physiological stimuli depending on the systemic hormonal environment and age [1]. The primary function of ligaments and tendons is to move from muscles to tendons or to assist movement to transfer force from the bone involved in the movement to the bone (ligaments). Foot and hand tendons net occur in relation to the ligament between them and this is called super-tendons. The concept of super-tendons has been proposed to explain that such networks exhibit a more functional range than their members [2].

In the organism, ligaments and tendons act as connective tissues that act as force-transmitting structures and provide musculoskeletal movement. Typical features of normal tendon tissue are parallel-aligned tenocytes and collagen I fibers. In addition, the extracellular matrix consists of proteoglycans, elastin, and glycoproteins. There is almost no vein in the tissue and nutrition is provided along with oxygen as well as nutrition at the osteotendinous junctions and vascularized myotendinous. Growth factors are vital for tendon homeostasis, development, and regeneration. The most important of these is growth factor-beta. Structural changes on tendinopathy and aging comprise the degree of vascularization (aging leads to less tendinopathy and more vascularization), extracellular matrix (age-related lower collagen content and tendinopathic collagen disorder), and proteoglycan tendon (small tendons, tendinopathic tendons) [3].

Tendons' basic structural properties situations are combined and shown in **Figure 1**. The main differences in morphology and organization of collagen fibers, has in terms of vascularization and cell density and morphology. In addition, extracellular matrix proteins in the normal aging and degenerative change the condition of the tendon and ligament [4]. Aging tendon tissue is different in terms of the tendon cells from healthy tissue morphology and finer turn into tenocytes have larger nuclei in older age. As for the vascularization is reduced and there are fat deposits in the connective tissue. Finally, tendinopathic tendon is more vascularized than normal tendon with irregular collagen fibers and the enriched with extracellular matrix proteoglycans (**Figure 2**).

According to the figure, healthy tendon tissue consists of densely packed collagen fibers in an amorphous ground material containing connective tissue of

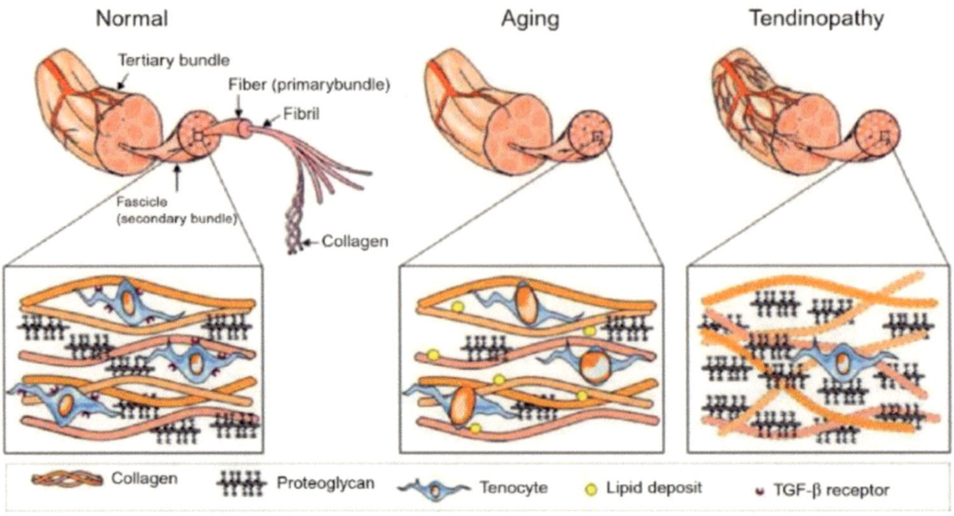

Figure 1.
Structural changes of tendons [4].

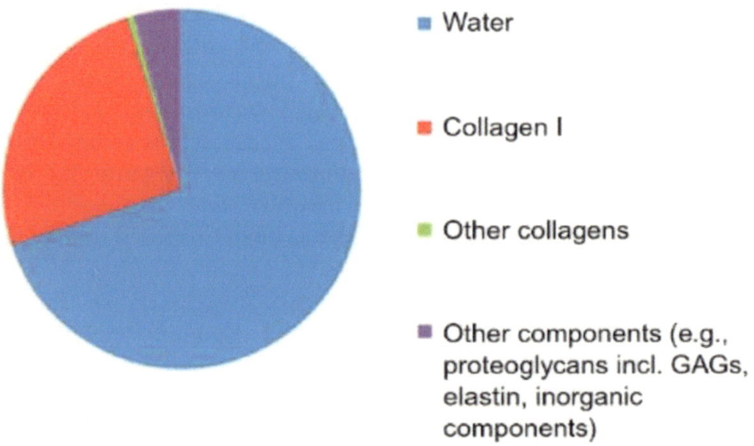

Figure 2.
Composition of tendon tissue [3].

water (60–80% of total wet weight), collagen (65–86% of dry weight, mostly type I collagen 95–98%) proteoglycans (1–5%), elastin (1–2%) and 0.2% inorganic components. In addition, tendon cells appear arranged in parallel lines [5].

The biomechanical behavior of a tendon is not only related to the magnitude of tension stress, but also to the shape of the tendon itself. The muscles used to perform precise and precise movements such as bending of the fingers have long and thin tendons, while those who perform strength and endurance actions such as quadriceps femoris and sural turaleps have shorter and more robust tendons. A short tendon has greater tensile strength than a long tendon because the load required to achieve fracture is much larger in the short tendon of the same diameter. A long tendon may undergo a greater deformation than a short tendon before it tears. The strength and resistance of a tendon are therefore two different entities and depend on the diameter and length of the tendon itself. The biomechanical properties of the tendon are related to the diameter and arrangement of collagen fibrils, tendons exposed to high stress are less flexible, large-diameter fibrils than small-diameter fibers [1].

The cells forming the tendons are generally thought to consist of tenocytes only for maintenance, repair, and regeneration. In scientific research, special cell types have been observed in tendons that can self-proliferate and differentiate into different cell types [6, 7]. In 2007, Bi et al. directly showed the presence of stem cells in tendons. Bi et al. showed that there is a small cell population carrying stem cell characters such as the ability to clone, self-proliferate and differentiate into other cells in human and mouse tendons [8]. After these developments, interest in tendon physiology, pathology, and tendon tissue engineering has increased. Tendon-derived stem cells, like other stem cells, play a role in tissue regeneration, maintenance, and repair within their local microenvironment. The in vivo niche environment of tendon-derived stem cells is still unknown. Although tendon stem cells are generally used to identify these cells, many names are used in the literature.

Muscles and tendons are the most commonly injured tissues in sports individuals. The limited number of studies on muscle-tendon injuries, especially in childhood and adolescence, has caused our knowledge to be quite limited compared to bone, growth cartilage, joint cartilage, and ligament injuries. Acute injuries such as muscle contusion or strain are seen in childhood and adolescence, mostly due to macro trauma. These injuries usually have a limited, benign course and allow the athlete to return to training and competitions in a short time. Overuse injuries resulting from repeated microtrauma and prolonged exposure to submaximal stress, although less common than acute injuries, require a more intensive treatment program. Overuse injuries cause individuals to stay away from sports for longer. Collagen connective tissue is an important part of a healthy tendon and in athletic performances, its robust function is a prerequisite for the smooth functioning of muscle-tendon units. An accurate understanding of the structure and metabolism of the tendon connective tissue is necessary to understand the etiology and pathogenesis of tendon injuries and athletic diseases and histopathological findings in each disease. In addition, this information is required to plan the treatment and rehabilitation protocol for a patient with a specific tendon problem. Future basic tendon science studies, not only in the field of rehabilitation and medicine but also in sports medicine, should explain where the pain after injury comes from chronic tendon disorders and how it can accelerate and accelerate tendon tissue healing after an injury [9].

Author details

Hasan Sözen
University of Ordu, Ordu, Turkey

*Address all correspondence to: hasansozen@odu.edu.tr

References

[1] Bordoni B, Varacallo M. Anatomy, Tendons. StatPearls, NCBI Bookshelf; 2018

[2] Benjamin R. The structure of tendons and ligaments. In: Archer C, Ralphs J, editors. Regenerative Medicine and Biomaterials for the Repair of Connective Tissues. Oxford/Cambridge/New Delhi: Woodhead Publishing Limited; 2010

[3] Buschmann J, Bürgisser GM. Biomechanics of Tendons and Ligaments. Duxford/Cambridge/Kidlington: Woodhead Publishing; 2017. Woodhead Publishing is an imprint of Elsevier

[4] Nourissat G, Houard X, Sellam J, Duprez D, Berenbaum F. Use of autologous growth factors in aging tendon and chronic tendinopathy. Frontiers in Bioscience. 2013;**E5**:911-921

[5] Kahn CJF, Dumas D, Arab-Tehrany E, Marie V, Tran N, Wang X, et al. Structural and mechanical multi-scale characterization of white New-Zealand rabbit achilles tendon. Journal of the Mechanical Behavior of Biomedical Materials. 2013;**26**:81-89

[6] Scutt N, Rolf CG, Scutt A. Glucocorticoids inhibit tenocyte proliferation and tendon progenitor cell recruitment. Journal of Orthopaedic Research. 2006;**24**(2):173-182

[7] Bi Y, Ehirchiou D, Kilts TM, Inkson CA, Embree MC, Sonoyama W, et al. Identification of tendon stem/progenitor cells and the role of the extracellular matrix in their niche. Nature Medicine. 2007;**13**(10):1219-1227

[8] Zhang J, Wang JHC. Characterization of differential properties of rabbit tendon stem cells and tenocytes. BMC Musculoskeletal Disorders. 2010;**11**:10

[9] Kannus P. Structure of the tendon connective tissue. Scandinavian Journal of Medicine and Science in Sports. 2000;**10**:312-320

Chapter 2

Tendon Structure and Classification

Murat Kaya, Nazım Karahan and Barış Yılmaz

Abstract

Tendons play an important role in the movement by transmitting the contraction force produced by the muscles to the bone they hold, and their contribution to stability to the joints is extremely important. Tendons generally have a very complex structure; they are actually heavily composed of connective tissue and have a small number of cells and rich extracellular matrix, similar to other connective tissue structures. The tendons are mainly composed of three parts: the tendon itself, the muscle-tendon junction, and the bone insertion. The simplest classification for the tendons classified according to their shapes, settlements, and anatomical structures is the classification made according to their shapes. Tendons can be classified in many ways according to their location, but the most logical one is the tendon classification in relation to the functions they see as intraarticular and extraarticular. According to their anatomy, the tendons can also be classified as sheathed or synovial-coated or unsealed or paratenon-coated. According to their functions, tendons can be classified as energy storage or positional tendons.

Keywords: tendon, tendon structure, tendon classification, fascicle, endotenon, epitenon, paratenon, collagen fibrils

1. Introduction

Tendons are dense fibrous tissues that bind the muscles to the bone. They play an important role in the movement by transmitting the contraction force produced by the muscles to the bone they hold. At the same time, their contribution to stability to the joints is extremely important. Although they differ in shape and size depending on the location, the common feature of all is that they can attach to a bone and transmit large loads without deforming them. Although they are structurally sound as they can withstand very high powers due to their function, degeneration and various damages caused by aging can result in loss of muscle strength [1–3].

Although tendons generally have a very complex structure, they are actually heavily composed of connective tissue and have a small number of cells and rich extracellular matrix, similar to other connective tissue structures. In terms of total tissue volume, while the cellular structure constitutes approximately 20% of total tissue volume, the remaining cells form 80% of the extracellular matrix. As a result of these factors, the cellular structure is mainly 60–85% collagen, 0.2% proteoglycans such as inorganic substances, 2% elastin, and 4.5% other proteins, while the matrix is composed of 55–70% water and the rest of the extracellular matrix consists of proteoglycans [4, 5].

2. Morphology, histology, microanatomy, and cell biology

When we look at the structure, tendons are composed of collagen fibrils; they consist of fiber bundles, fascicles, and finally the tendon structure, also known as a group of fascicles. In conclusion, tendons are composed of multiple bundles, fibroblast, and dense linear collagen fibrils, which form the macroscopic structure of tendons and give the appearance of fibrous. In general, connective tissue surrounding the tendons allows some friction. In this way, the ligament around many tendons has a mesotendon that sticks to the tissue and encircles it. This structure also allows the tendon to flush. The connective tissue of low density surrounds tendon fascicles, which is called the endotendon. The fact that tendon fascicles are surrounded by endotendon actually allows tendon bundles to make small slip motion. Endotendon tissue continues in the form of an epitendon covering the tendon surface. When the tendon joins with the muscle, it continues as epimysium in the epitendon muscle. At this point, the muscle-tendon junction must transmit the muscle contraction to the tendon exactly. The tendon adhesion of the muscle occurs when the fibrous tissue layers of the muscle enter the collagen fibers of the tendon into the collagen fibers. In a study conducted by electron microscopy, the position of the muscle cells and tendons is like the fingers of two hands that are locked together. Collagen fibers do not enter the muscle cells, but they bind tightly under the basal membrane. The movement of a normal tendon, the transfer of muscle power for the entire movement of the joints, and the feeding of tendons depend on peritendinous connective tissue. This structure is called the peritendon. These structures form the sheaths, which are very finely organized structures from the loose connective tissue [3, 6].

The cell and matrix compositions of tendons are similar to ligaments and capsules and contain only small differences. In fact, they all have the same cell type and similar vascular and innervation sources. Collagen, elastin, proteoglycan, and noncollagenous proteins combine to form the macromolecular framework of dense fibrous tissues. In all of them, the dominant cell type is fibroblasts. In particular, the cells within the tendons are specific fibroblasts called tenocytes. The main role of these cells is to control cell metabolism (production and degradation of extracellular matrix) and to react to mechanical stimuli applied to the tendon. Especially tensile loads act as a signal for collagen production, and this process is called mechanical transmission. These cells stretch along collagen fibrils in the form of longitudinal arrays where they have a tensile load [7, 8].

The extracellular matrix of tendons is largely composed of collagen fiber network and less proteoglycans, elastin, and other proteins. The main task of these components is to maintain the structure of the tendon and facilitate the biomechanical reaction of the tissue against mechanical loads. An important component of extracellular matrix, proteoglycans, forms less than 1% of the dry weight [9].

The main substance in tendons and ligaments is basically about 0.2% inorganic substances and about 4.5% other proteins. The most effective of inorganic substances are proteoglycans. In addition to prostaglandins with a small amount in the main substance, the most common biomechanical properties are the decorin and cartilage oligomeric matrix protein (COMP) [10].

The protein clusters in the structure are connected to a large portion of the extracellular matrix of tendons, making the matrix a structure similar to the gel. Thanks to this compound, collagen provides spaces and lubrication between microfibrils, while cement-like material also makes the collagen structure of tendons stable and contributes to the resistance of the tissue [3, 11, 12].

The collagen in the tendon structure is found as the main molecule of dense fibrous tissue and forms approximately 70% of dry weight. When examined as collagen type, it is largely composed of Type I (60%) and other types, namely, Types III, IV, V, and VI. Collagen Type-I fibers are capable of withstanding large tensile loads and are found in abundance from the tendon structure, allowing a certain degree of stretch and mechanical deformations of the tendons [13].

According to today's information, synthesis of collagen in connective tissue begins in the cell membrane of fibroblasts. This synthesis process is similar to that of all connective tissue, although it may differ slightly depending on the type of complex collagen. Therefore, tendons, which contain Type-I collagen, have a process of synthesis and degradation similar to those in the ligaments and bones. From here, with a more detailed look, we can say that synthesizing for collagens in tendon structure begins in the cell membrane of the tenocytes. "Integrin" molecules have an important role in collagen production because they are sensitive to the transmission of mechanical charge from inside the cell to the outside or vice versa. In other words, the integrins are like force sensors and, in particular, detect cell withdrawal, allowing the cell to react to these mechanical stimuli. At the same time, various growth factors contribute to the regulation of this mechanical conversion process [14].

Cross-linkages form between collagen molecules, which are very important for clustering at the fibril level. The cross-links between the fibrils are more complex. And this cross-link structure of collagen fibrils provides the strength of the tissue and thus ensures that it performs the task of the tissue under mechanical loads. In the newly formed collagen, these cross bonds are less in number, soluble in salt or acid solution, and can easily break with heat. As collagen matures, the number of cross bonds that can dissolve and break down decreases and decreases to the minimum level. As a result, organized collagen molecules form microfibril, sub-fibrils, and fibrils. The fibrils are also clustered to form collagen fibers, collagen clusters or fascicles, and the tendon. Tenocytes are arranged between these fascicles and aligned in the direction of the mechanical load [10].

In the cellular structures of tendons, as mentioned above, there is much less amount of elastin than collagen, because the mechanical properties of the tendons depend not only on the architecture and properties of collagen fibers but also on the extent to which this structure contains elastin. However, in tendons, elastin proteins, which usually constitute about 2% of the dry weight, can be up to 70% in elastic bonds such as nuchal ligament and ligamentum flavor. Because the bond has a special function and the nerve roots of the spine, mechanical stresses, stresses, etc. provide stability to the spine [9, 15].

Blood circulation in tendons is very important, because the current circulation of blood directly affects metabolic activity especially during healing. However, blood circulation in tendons is not as rich as muscles and bones, and it accounts for only 1–2% of the extracellular matrix. Therefore, they have a white color when compared to the muscles with a much higher blood vessel density. However, there are a few factors such as the anatomical location, structure, previously damaged condition, and physical activity level of tendons that contribute to blood supply besides the small amount of vascular structure. There are studies that show that blood flow increases in tendons in the case of increasing physical activity in the literature. There are more vascular tendons due to their anatomical position or shape and function. The flushing of tendons is primarily derived from the synovium at the point of attachment to the bone or paratenon. However, some tendons feed on the tendon like the Achilles tendon and the paratenon structure, and some tendons are fed by a true synovial sheath they are surrounded. Bone and tendon adhesion is a layer of cartilage where blood flow cannot pass directly from the bone-tendon compound. Instead, they make anastomosis with the veins on the periosteum and make indirect connections [16].

In contrast, tendons have a very rich neural network and are often innervated from the muscles in which they are associated or from the local cuticle nerves. However, experimental studies on humans and animals have shown that tendons have different characteristics of nerve endings and mechanoreceptors. They play an important role especially for proprioception (position perception) and nociception (pain perception) in joints. In fact, studies have shown that there is internal growth in the nervous and vascular systems during the healing of tendon, which causes chronic pain. Internal growth of the vein is an indicator of the tendon trying to heal, but because of this growth, nerves may feel pain in areas without pain before. This means that the nerves play an important role not only in the proprioception but also in the nociception. Nerve endings are located below the muscle-tendon junction and typically in the bone-tendon junction in the form of Golgi organs, Pacini bodies, and Ruffini endings. Of these, the Golgi organs are only mechanically stimulated by pressure and compression, so that they receive information from the power produced by the muscle. Pacinian bodies are rapidly adaptive mechanoreceptors due to nerve endings with a highly sensitive capsular end to deformation, thus dynamically responding to deformation, but are insensitive to constant or stable changes. Ruffin termination results from multiple, thin capsule-tipped, and single axons and has slowly adapting mechanoreceptors and thus continues to receive information until a constant warning level is stimulated during deformation [17].

The tendons are surrounded by loose, porous connective tissue, which is called paratenon. A complex structure, paratenon, protects the tendon and allows shifting tendon cover format. Tendon sheaths consist of two continuous layers: parietal on the outside and visceral on the inside. The visceral layer is surrounded by synovial cells and produces synovial fluid. In some tendons, the tendon sheath extends along the tendon, while in others it is found only in the binding parts of the bone.

The parietal synovial layer is found only under the paratenon in the body regions where tendons are exposed to high friction. This is called the epitenon and surrounds the fascicles. In this case, epitenon's synovial cells produce lubricating liquid. In regions where friction is less, tendon is surrounded by paratenon only. At the tendon-bone junction, the collagen fibers of endotenon continue into the bone and become a peritendon.

The regions of the tendon bonding to the bone consist of a dense connective tissue, which is able to adhere to the hard bone from the dense connective tissue and is resistant to movement and damage. Although they occupy a small area in size, the areas of adhesion to the bone have a complex structure that is much different from that of the tendon itself. According to the size of the load they carry, they show a different proportion of collagen bundles [18].

The tendons cling to the bone is a complex event; collagen fibers mix into fibrocartilage, mineralize, and then merge with the bone. "Sharpey's penetrating fibers" continue with the external lamellar structure of the bone of tendon fibrosis along the period that is important for the entry of the tendon called enthesis. Sticking to the bone is done in two ways. In the first type, the adhesion of many collagen fibers is direct to the bone, while the second type indirectly adheres to the periosteum. In other words, the tendon is attached to the bone in the form of fibrous or indirect adhesion to the metaphysics and diaphysis of long bones or fibrocartilaginous or direct adhesion to the epiphyses of the bone. In fibrous adhesions, while the collagen fibers of the tendon are permanently adhered to the periosteum during bone development, fibrocartilaginous adhesions have a gradual transition from tendon to bone. This gradual transition in fibrocartilaginous adhesions includes the tendon, decalcified fibrocartilage, calcified fibrocartilage, and four zones of bone, so that the uniform distribution of the load at the adhesion site and the joint movement and the coordination of the collagen fibers are ensured. However, changes in the fibrocartilaginous structure due to compressive loading vary depending on the adhesion sites of the tendons. This ensures better protection against compressive forces. The bones of the tendons are composed of four regions within the bone; at the end of the tendon (region 1), collagen fibers enter the fibrocartilage (fibrous cartilage—region 2). As the fibrocartilage progresses, it becomes mineral fibrocartilage (area 3) and then integrates with cortical bone (fourth region). This transformation, which is more bone structure than tendon structure, leads to gradual increase of mechanical properties of the tissue [3, 19–21].

3. Classification of tendons

The tendons are mainly composed of three parts: the tendon itself, the muscle-tendon junction, and the bone insertion. In general, they pass through the joints and adhere to their distal. In this way, they increase the effectiveness of the muscles on the joints. At the same time, similar to bones, mechanical properties vary depending on the load carrying place. For this reason, knowing where they are helps us understand the structure. In fact, not every muscle has a tendon. While some tendons are involved in some muscles that play an active role in joint movements, the presence of some tendons is to increase muscle movement distances rather than the movement of the joint. For example, Achilles tendon is a very special tendon for the body carrying the loads by centralizing the strength of a few muscles. In contrast, some tendons, such as the posterior tibial tendon, act by distributing the load to several bones. Although it is known that most tendons originate from the muscle and adhere to the bone, some tendons may be the starting point for muscles, or two muscles are connected to each other through a tendon [22, 23].

The simplest classification for the tendons classified according to their shapes, settlements, and anatomical structures is the classification made according to their shapes. They can be very small and very long, and they can be very large and very short. Tendons are very variable according to their shape, long, round, rope-shaped (such as Achilles tendon), or short; flat tissue adhesion (such as bicipital aponeurosis) can be seen. In other words, tendons may change from flat to cylinder, from fan shape to ribbon shape. However, round tendons (such as flexor digitorum profundus) or flat tendons (such as rotator cuff, bicipital aponeurosis) are more involved in the body. In this simple classification, tendons are divided into round and flat and are very different from each other as structural and functional. For example, while round tendons respond equally to tensile loads with parallel collagen patterns, flat tendons such as rotator cuffs can respond microanatomically in the form of compression and shear forces due to longitudinal, oblique, and transverse collagen sequences. However, in round tendons, the section area is proportional to the maximum isometric strength of the muscle. In other words, due to parallel collagen sequences, flat tendons are resistant to compression and shear forces due to flat, longitudinal, and oblique collagen sequences in comparison to round tendons that respond equally to the tensils [3, 24].

Tendons can be classified in many ways according to their location, but the most logical one is the tendon classification in relation to the functions they see as the intraarticular (biceps long head and popliteus tendon) and the extraarticular (Achilles tendon). Most tendons are non-articular, but the intra-articular ones lack the ability to repair after injury as in the same intra-articular ligaments (an example of anterior cruciate ligament tear). At the same time, although most tendons adhere to the bone, some tendons form the origo point for the muscles (lumbrical muscles originate from the flexor digitorum profundus) or connect two muscles (such as

	Energy storage tendons	Positional tendons
Function	-Storage and release of elastic stress energy	-Transport the forces created in muscles to the bones
Material specifications	-Bimodal with smaller fiber diameter -More glycosaminoglycan and water content, softer matrix -Increased interfascicular slip due to lower intrafascicular rigidity	-Unimodal with a wider diameter of a fiber -Lower glycosaminoglycan and water content, the harder matrix -Tightly packed fascicles with less interfascicular slip at low loads
Biomechanical features	-It can extend in physiological loads -Higher tensile strength -Lower tensile strength	-Cannot stretch in physiological loads -Lower tensile strength -Higher tensile strength
Injury	-More	-Less
Example	-Achilles tendon	-Anterior tibial tendon

Table 1.
Classification and properties of tendons according to their functions.

omohyoid and digastric muscle). In addition, the large part of the tendon may originate from the muscle itself (gastrocnemius and soleus). For example, in some muscles tendons move into the muscle joint and tendon sticks at an angle. This allows a high proportion of muscle fibers to adhere to the tendon, thereby increasing the strength of the muscle-tendon unit but reducing the range of motion.

According to their anatomy, the tendons can also be classified as sheathed or synovial-coated (such as the long flexor of the fingers) or unsealed or paratenon-coated (such as Achilles tendon). In other words, these tendons, which are separated by intrasynovial and extrasynovial, have a higher slippage resistance compared to the intrasynovial tendon structure, when examined more closely. At the same time, the soft tissue protection and vascularity of these two tendons are different [20].

According to its functions, tendons can be classified as energy storage or positional tendons (**Table 1**). In general, the muscles tend to tendon to shorten the stress load; the affected tendon is stretched and the muscle can relax again when relaxed. This makes the tendon a structure that stores elastic voltage energy. The best example of energy storage tendons is Achilles tendon. Tibialis anterior tendons in human are examples of positional tendons, and they can never extend relatively. Positional tendons are rarely injured because they extend less [25–27].

4. Conclusion

In conclusion, tendons are composed of multiple bundles, fibroblast, and dense linear collagen fibrils, which form the macroscopic structure of tendons and give the fibrous appearance. The cell and matrix compositions of tendons are similar to ligaments and capsules and contain only small differences. In fact, they all have the same cell type and similar vascular and innervation sources. The extracellular matrix of tendons is largely composed of collagen fiber network and less proteoglycans, elastin, and other proteins. The main task of these components is to maintain the structure of the tendon and facilitate the biomechanical reaction of the tissue against mechanical loads.

Knowing where tendons are helps us understand the structure. While some tendons are involved in some muscles that play an active role in joint movements, the presence of some tendons is to increase muscle movement distances rather than the movement of the joint.

DOI: http://dx.doi.org/10.5772/intechopen.84622

Author details

Murat Kaya[1]*, Nazım Karahan[2] and Barış Yılmaz[1]

1 Department of Orthopedic Surgery and Traumatology, Fatih Sultan Mehmet Education and Research Hospital, Istanbul, Turkey

2 Department of Orthopedic Surgery and Traumatology, Selahaddin Eyyubi State Hospital, Diyarbakır, Turkey

*Address all correspondence to: kayamuratdr@gmail.com

References

[1] Frank F, Shrive NG. Molecular and biomechanics of normal and healing ligaments: Review. Osteoarthritis and Cartilage. 1999;**7**:130-140

[2] Gitto S, Draghi F. Normal sonographic anatomy of the wrist with emphasis on assessment of tendons, nerves, and ligaments. Journal of Ultrasound in Medicine. 2016;**35**(5):1081-1094

[3] Heybeli N, Komur B, Yılmaz B, Guler O. Tendons and ligaments. In: Korkusuz F, editor. Musculoskeletal Research and Basic Science. London: Springer; 2016. pp. 4465-4482

[4] Nakamura N, Hart DA, Boorman RS, Kaneda Y, Shrive NG, Marchuk LL, et al. Decorin antisense gene therapy improves functional healing of early rabbit ligament scar with enhanced collagen fibrillogenesis in vivo. Journal of Orthopaedic Research. 2000;**18**:517-523

[5] O'Brien M. Anatomy of tendons. In: Maffuli N, Renstrom P, Leadbetter WB, editors. Tendon Injuries. London: Springer; 2005. pp. 3-13

[6] Benjamin M, Ralphs JR. The cell and biology of tendons and ligaments. International Review of Cytology. 2000;**196**:85-130

[7] Winters SC, Seiler JG 3rd, Woo SL, Gelberman RH. Suture methods for flexor tendon repair: A biomechanical analysis during the first six weeks. Annales de Chirurgie de la Main et du Membre Supérieur. 1997;**16**:229-234

[8] Lo IK, Chi S, Ivie T, Frank CB, Rattner JB. The cellular matrix: A feature of tensile bearing dense soft connective tissues. Histology and Histopathology. 2002;**17**:523-537

[9] Majima T, Marchuk LL, Sciore P, Shrive NG, Frank CB, Hart DA. Compressive compared with tensile loading of medial collateral ligament scar in vitro uniquely influences mRNA levels for aggrecan, collagen type II and collagenase. Journal of Orthopaedic Research. 2000;**18**:524-531

[10] September AV, Schwellnus MP, Collins M. Tendon and ligament injuries: The genetic component. British Journal of Sports Medicine. 2007;**41**(4):241-246

[11] Tresoldi I, Oliva F, Benvenuto M, Fantini M, Masuelli L, Bei R, et al. Tendon's ultrastructure. Muscle, Ligaments and Tendons Journal. 2013;**3**(1):2-6

[12] Wang JH, Guo Q, Li B. Tendon biomechanics and mechanobiology: A mini review of basic concepts and recent advancements. Hand Therapy. 2012;**25**(2):133-140

[13] Thorpe CT, Birch HL, Clegg PD, Screen HR. The role of the non-collagenous matrix in tendon function. International Journal of Experimental Pathology. 2013;**94**(4):248-259

[14] Reuther KE, Gray CF, Soslowsky LJ. Form and function of tendon and ligament. In: O'Keefe RJ, Jacobs JJ, Chu CE, Einhorn TA, editors. Orthopaedic Basic Science. 4th ed. Rosemont: American Academy of Orthopaedic Surgeons; 2013. pp. 213-228

[15] Doral MN, Alam M, Bozkurt M, Turhan E, Atay OA, Donmez G, et al. Functional anatomy of the Achilles tendon. Knee Surgery, Sports Traumatology, Arthroscopy. 2010;**18**(5):638-643

[16] Peacock EE. A study of circulation in normal tendons and healing grafts. Annals of Surgery. 1959;**149**:415-428

[17] Ackermann PW. Neuronal regulation of tendon homoeostasis.

International Journal of Experimental Pathology. 2013;**94**(4):271-286

[18] Meimandi-Parizi A, Oryan A, Moshiri A. Role of tissue engineered collagen based tridimensional implant on the healing response of the experimentally induced large Achilles tendon defect model in rabbits: A long term study with high clinical relevance. Journal of Biomedical Science. 2013;**20**(1):28

[19] Blevins FT, Djurasovic M, Flatow EL, Vogel KG. Biology of the rotator cuff tendon. The Orthopedic Clinics of North America. 1997;**28**:1-16

[20] Reddy GK, Stehno-Bittel L, Enwemeka CS. Matrix remodeling in healing rabbit Achilles tendon. Wound Repair and Regeneration. 1999;**7**:518-527

[21] Murray MM, Spector M. The migration of cells from the anterior cruciate ligament into collagen glycosaminoglycan regeneration templates in vitro. Biomaterials. 2001;**22**:2393-2402

[22] Woo SLY, An KN, Frank CB. Anatomy, biology, and biomechanics of tendon and ligament. In: Buckwalter TA, Einhorn TA, Simon SR, editors. Orthopaedic Basic Science: Biology and Biomechanics of the Musculoskeletal System. 2nd ed. Rosemont, IL: American Academy of Orthopaedic Surgeons; 2000. pp. 581-616

[23] Vesentini S, Redaelli A, Gautieri A. Nanomechanics of collagen microfibrils. Muscle, Ligaments and Tendons Journal. 2013;**21**(1):23-34

[24] Khan U, Kakar S, Akali A, Bentley G, McGrouther DA. Modulation of the formation of adhesions the healing of injured tendons. The Journal of Bone and Joint Surgery. British Volume. 2000;**82**:1054-1058

[25] Thorpe CT, Udeze CP, Birch HL, Clegg PD, Screen HR. Specialization of tendon mechanical properties results from interfascicular differences. Journal of the Royal Society Interface. 2012;**9**(76):3108-3117

[26] Batson EL, Paramour RJ, Smith TJ, Birch HL, Patterson-Kane JC, Goodship AE. Are the material properties and matrix composition of equine flexor and extensor tendons determined by their functions? Equine Veterinary Journal 2003;**35**(3):314-8

[27] Birch HL. Tendon matrix composition and turnover in relation to functional requirements. International Journal of Experimental Pathology. 2007;**88**(4):241-248

Chapter 3

Imaging of Tendons

Stephanie C. Torres-Ayala, Alvaro Bravo-Martínez, Amanda P. Marrero-González, Luis R. Rodriguez-Ortíz and Jorge A. Vidal

Abstract

Magnetic resonance imaging (MRI) and ultrasound (US) are useful radiologic modalities that allow adequate evaluation of tendon anatomy and integrity. Each modality contains unique advantages as diagnostic tools, allowing detection of tendon injuries and pathology. This chapter focuses on the key imaging features of tendons in both ultrasound and magnetic resonance, with emphasis on the major joints such as the shoulder, elbow, hand/wrist, hip, knee and foot/ankle joints. Each section provides a review of standard magnetic resonance imaging protocols and ultrasound technique, along with a discussion of the radiologic appearance of the most common tendon pathology affecting each joint.

Keywords: tendons, ultrasound, magnetic resonance imaging, radiology

1. Introduction

Imaging modalities play a significant role in the evaluation of tendon pathology. Knowledge of anatomical landmarks of the tendons is of utmost importance in order to differentiate and accurately diagnose pathologic processes. Ultrasound (US) and Magnetic Resonance Imaging (MRI) are currently the imaging modalities of choice to evaluate tendon pathology, each with its own unique advantages as diagnostic tools. Tendons are characteristically seen as echogenic fibrillar structures on US and as homogeneous hypointense structures on MRI.

This chapter focuses on the key imaging features of tendons in US and MRI modalities, with emphasis on the shoulder, elbow, hand, hip, knee and foot joints. Each chapter section will provide a review of the anatomical landmarks, normal US and MRI appearance, imaging protocols, and discussion of common pathology affecting the tendons in each joint.

2. Shoulder

2.1 MRI protocol

The patient is positioned supine with the arm at the side in neutral position or slight external rotation in order to put some tension on the long head of the biceps tendon. A small field of view (approximately 14–16 cm) is obtained in three imaging planes: axial, coronal oblique, and sagittal oblique. The axial images are acquired from the top of the acromioclavicular joint through the proximal humeral shaft including the insertion of the pectoralis muscle. The coronal oblique images are obtained with planes made parallel to the supraspinatus tendon or in a plane perpendicular to the articular surface of the glenoid, ranging from the coracoid process to the infraspinatus muscle. Finally, the sagittal oblique images are acquired with planes parallel to the articular surface of the glenoid, from the scapular neck through the lateral aspect of the humerus [1]. A standard shoulder MRI usually includes sagittal oblique T1-weighted image (T1WI), fast spin echo (FSE) T2-weighted image (T2WI) with fat suppression, coronal oblique FSE T2WI with fat suppression, and axial FSE T2WI and FSE proton density (PD) with fat suppression.

2.2 Ultrasound examination technique

Sonographic evaluation of the shoulder can be performed with the following steps [2]:

1. Long head of biceps brachii tendon: The patient places the hand on his or her lap, as this position rotates the bicipital groove anteriorly. The transducer is placed in the axial plane over the anterior aspect of the shoulder to identify the bicipital groove, where the long head of the biceps brachii tendon is found. The long head of the biceps brachii tendon is followed proximally to where the bicipital groove becomes shallow and then distal to the level of the pectoralis major tendon. The transducer is then turned 90° to visualize the tendon in long axis from the humeral head to the pectoralis tendon.

2. Subscapularis tendon: The transducer is placed in the axial plane, as in the previous step, to first visualize the bicipital groove and then centered over the lesser tuberosity at the medial aspect of the bicipital groove. Then the patient rotates the shoulder externally to pull the subscapularis tendon laterally, which will orient the tendon fibers perpendicular to the transducer sound beam and eliminate anisotropy. Then it may be moved laterally over the bicipital groove to ensure that the long head of the biceps brachii tendon is within the bicipital groove, and rule out subluxation or dislocation, which may be present only in external rotation [3]. The subscapularis tendon can also be evaluated in short axis by turning the transducer 90°.

3. Supraspinatus and infraspinatus tendon: One way to evaluate the supraspinatus tendon is to ask the patient to place the dorsum of his or her ipsilateral hand behind the back, called the Crass position [4]. This position pulls the tendon out from under the acromion. The Crass position is very helpful in localizing the greater tuberosity but its limitations include poor visualization of the rotator interval and patient discomfort [3]. Because of these disadvantages the modified Crass position is more commonly used, by asking the patient to place his or her ipsilateral hand on the hip or buttock region. This position places the greater tuberosity more lateral than with the Crass position, and also allows easy visualization of the rotator interval with little patient discomfort [5]. Either in the Crass or modified Crass position, supraspinatus evaluation should begin by observing the tendon in long axis as this allows visualization of the three surfaces (articular, bursal, greater tuberosity) [6]. Scanning should be continued anteriorly along the greater tuberosity until the intra-articular portion of the biceps tendon is identified. The infraspinatus tendon is evaluated by moving the transducer posteriorly over the middle facet of the greater tuberosity.

2.3 Supraspinatus tendon

The supraspinatus tendon arises from the supraspinous fossa, runs between the undersurface of the acromion and the top of the humeral head, and inserts into the most superior facet of the greater tuberosity of the humerus. On MRI, the entire length of the supraspinatus tendon can be seen well in the coronal oblique plain, running at an angle of approximately 45° [7]. The musculotendinous junction of the tendon normally is located just lateral to the acromioclavicular joint. On sagittal oblique images, the supraspinatus tendon is imaged in cross section, which is valuable to confirm the status of the tendon when abnormalities are seen in the plane of imaging, where the tendon is viewed longitudinally.

The normal sonographic appearance of the supraspinatus tendon is hyperechoic and fibrillar with a convex superior margin at the level of the superior facet of the greater tuberosity of the humerus [8]. It parallels the curved contour of the humeral head, flattening out as it inserts into the greater tuberosity. The subacromial-subdeltoid bursa should be seen as a single thin hyperechoic line paralleling the tendon superiorly.

The supraspinatus tendon is the most commonly affected when compared to the other tendons of the shoulder [8]. There are multiple pathologies that may limit the space within the coracoacromial arch, producing impingement of this tendon. Abnormalities from impingement range from tendon degeneration to partial-thickness or full-thickness tears. Most partial-thickness tears occur in the articular aspect of the tendon, rather than on the bursal surface. Tears are usually located distally, either near its attachment to the greater tuberosity or in the critical zone located approximately 1 cm proximal to its insertion, and start in the anterior portion as rim rent tears and spread posteriorly [7]. Rim rent tears refer to disruption of the insertional fibers on the greater tuberosity. Complete disruption of the fibers with communication between the joint and the overlying bursa indicates a full-thickness tear (**Figure 1**).

Tendon degeneration usually demonstrates increased signal intensity on T1WI and T2WI, although not as high signal as fluid. However, a partial thickness tear demonstrates increased signal intensity on T2WI similar to fluid. Indications of a full thickness tear include: tendon discontinuity, fluid signal in tendon gap, and retraction of musculotendinous junction [7].

Tears in ultrasound are demonstrated as anechoic or hypoechoic defects, although acute tears will more likely appear anechoic like fluid [8]. As a supraspinatus tendon tear enlarges, tendon retraction and volume loss occur, with loss of the normal superior convex shape. The length or degree of retraction of a full thickness tear can be measured on longitudinal views oriented parallel to the long axis of the cuff and the width can be measured on transverse views oriented perpendicular to the long axis of the cuff [2]. On the other hand, tendinosis is usually less defined, and may be associated with increased tendon thickness, and not usually associated with adjacent cortical irregularity of the greater tuberosity.

2.4 Long head of biceps brachii tendon

The long head of biceps brachii tendon originates from the supraglenoid tubercle of the scapula, courses intra-articularly to the entrance of the bicipital groove and continues caudally, inserting along the radial tuberosity of the proximal radius. On MRI, portions of this tendon can be evaluated on coronal oblique images, from its origin at the superior labrum and inferiorly in the bicipital groove. The portion that is located within the bicipital groove is seen on axial images as a round or oval structure, and sometimes it may blend with the low signal intensity cortex of the humerus, making it difficult to identify. It is normal to find a small amount of fluid

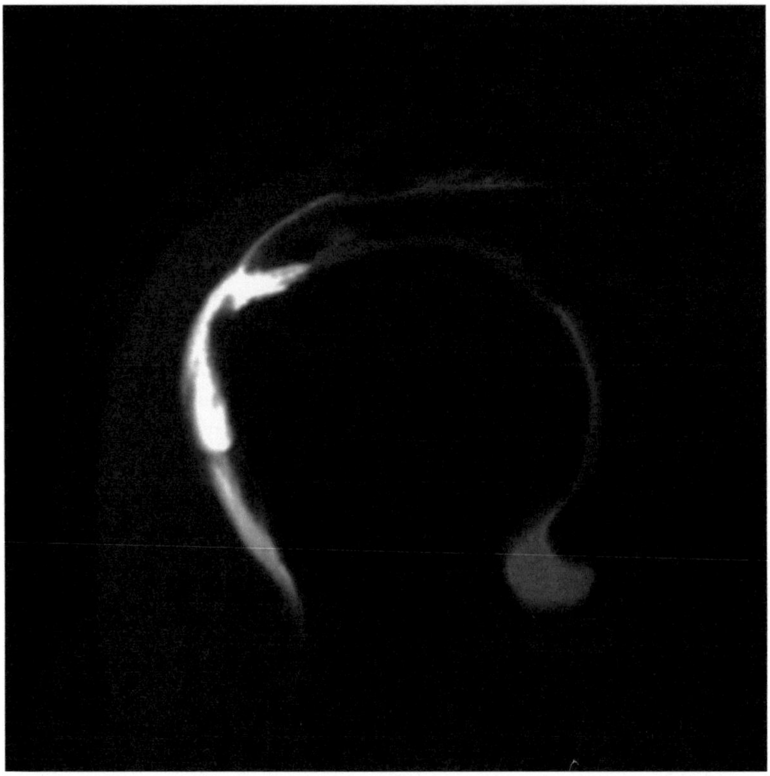

Figure 1.
Magnetic resonance arthrogram T1 fat saturated coronal oblique image shows a full thickness tear of the supraspinatus tendon with contrast leaking from the joint capsule into the subdeltoid space.

in the dependent side of the long head of the biceps tendon sheath, as the tendon sheath normally communicates with the glenohumeral joint. High signal round structures found lateral to the tendon within the bicipital groove represent the anterior circumflex humeral artery/vein and should not be confused with tenosynovitis [7].

The long head of biceps brachii tendon should be found within the intertubercular groove upon sonographic evaluation of the shoulder. The tendon fibers should be seen without tears, heterogeneity or thickening. The normal tendon will appear hyperechoic; however, because the tendon courses deep, it may appear artifactually hypoechoic due to anisotropy [8]. Adjusting the transducer to aim the sound beam perpendicular to the tendon fibers can eliminate this artifact.

The proximal aspect of the tendon may be affected by impingement in the same ways as the supraspinatus tendon because of its similar location and course beneath the supraspinatus tendon. Tears associated with impingement usually occur proximal to the bicipital groove and are usually seen in the older population. Acute tears unrelated to impingement are commonly secondary to a traumatic injury in young individuals, and usually occur distally in the tendon, near the musculotendinous junction [7]. When a full-thickness tear occurs, axial MRI images of the shoulder may show an empty bicipital groove, without evidence of the oval, low signal long head of the biceps tendon. An empty bicipital groove may also indicate tendon dislocation, which is also associated with disruption of the transverse humeral ligament that holds the biceps tendon in place. In this case, the low signal round tendon is seen medial to the bicipital groove, either deep or superficial to the subscapularis tendon, which usually also tears as well.

When shoulder effusion is present, fluid may be seen sonographically surrounding the biceps tendon at the level of the bicipital groove given the normal communication between the tendon sheath and joint. Joint effusion appears anechoic, however if fluid is complex it may be hypoechoic, isoechoic or hyperechoic relative to muscle, resembling synovial hypertrophy [8]. Tenosynovitis is favored over joint fluid extending into the sheath if there is focal distention of the tendon sheath with hyperemia and it is symptomatic with transducer pressure. Tendinosis of the long head of the biceps brachii should be considered when the tendon is abnormally hypoechoic and increased in thickness with lack of fiber disruption. Anechoic clefts or surface irregularity of the tendon favor a partial-thickness tear [9]. The primary finding in a full-thickness tear is lack of visualization of the biceps tendon or empty bicipital groove. As mentioned before, one must also consider tendon subluxation or dislocation when encountered with an empty bicipital groove. In this case, the tendon can be seen medially displaced, usually superficial to the lesser tuberosity.

3. Elbow

3.1 MRI protocol

Patients should be supine in a comfortable position with the arm to be imaged in supine position as well. Images should include from the distal humeral metaphysis up to the radial tuberosity and this area should be imaged in axial, coronal and sagittal planes. Sequences should include non-fat saturated T1, PD, and fat-saturated T2WI; gradient echo (GRE) may also be included depending on the pathology suspected [10].

3.2 Ultrasound examination technique

Ultrasound of the elbow usually focuses on the area of clinical interest, nonetheless, the anterior, lateral, medial and posterior compartments should all be evaluated. A high frequency linear transducer of 12–17 mHZ is preferred. To evaluate the anterior compartment of the elbow, which includes the distal biceps tendon, it should be extended with a supine forearm. Evaluation should include transverse and longitudinal planes from 5 cm proximal and distal to the joint. The lateral elbow compartment, which includes the common extensor tendon, is evaluated with the arm placed in internal rotation and elbow joint in flexion. The medial compartment includes the common flexor tendons, which is evaluated sonographically by extending the forearm in forceful external rotation. Lastly, the posterior elbow, which contains the distal triceps tendon, is evaluated by placing the elbow in 90° flexion with the arm internally rotated [11].

3.3 Common extensor tendon

The common extensor tendon attaches to the humeral lateral epicondyle uniting the individual tendons of the extensor carpi radialis brevis, extensor digitorum, extensor digiti minimi and the extensor carpi ulnaris. Normally the common extensor tendon is a band of low signal intensity on both T1WI and T2WI, seen superficial to the radial collateral ligament complex and the tendon should show complete fibers at its insertion in the lateral epicondyle.

A common cause for elbow pain is lateral epicondylitis, also known as tennis elbow. In these cases, the tendon may appear thickened with increased intermediate signal intensity on T1WI and T2WI. Abnormal fluid signal intensity may be seen traversing the tendon fibers in partial tendon tears, most common in the extensor carpi radialis brevis tendon [12] (**Figure 2**). If there is a fluid signal intensity gap with discontinuity of the tendon fibers, a full thickness tear is present. Avulsion injuries may be present when there is associated bone marrow edema at the tendinous insertion site. The US evaluation of lateral epicondylitis shows a heterogeneous tendon with focal hypoechoic areas.

3.4 Distal biceps tendon

The distal biceps brachii tendon is located at the anterior elbow compartment, coursing through the antecubital fossa with its distal insertion at the radial bicipital tuberosity. Its superficial fibers form the lacertus fibrosis which course medially to form the distal portion of the tendon. Pathology of the distal biceps tendons is most common in people who perform heavy weightlifting, with increased risk in those who use anabolic steroids [12]. A distal biceps tendon tear results in retraction of the myotendinous junction, clinically known as a Popeye's sign or mass in the proximal arm. This is seen as complete discontinuity of the tendon fibers, best appreciated in axial and sagittal planes. In order to be able

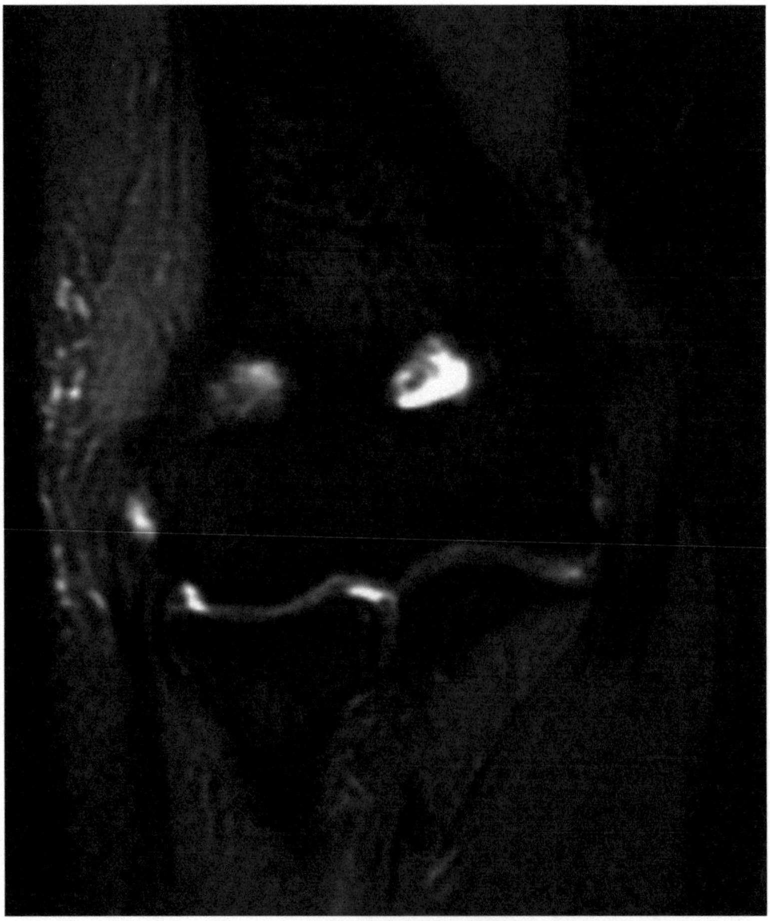

Figure 2.
Proton density fat saturated coronal image of the elbow shows fluid signal at the insertion of the common extensor tendon consistent with a tendon tear.

to visualize the retracted tendon and area of avulsion at its distal insertion the arm may be supine, flexed and abducted. If there is a partial tear present, then on magnetic resonance imaging there will be peritendinous increased T2 signal intensity [12].

4. Hand/wrist

4.1 MRI protocol

For MR imaging of the hand the patient is placed in prone position, with the arm elevated above the head, also known as the "superman position". When specifically imaging the thumb, the latter should be fully extended and at the center of the scanner and foam pads may be used for fixation of the area of interest. Small surface or dedicated hand or wrist coils are important in order to obtain high quality images. Axial images with respect to the fingers are first obtained and these are then used to plan sagittal and coronal views. When imaging the thumbs, coronal and sagittal views should be tilted 90° to sesamoids at the level of the metacarpophalangeal joint (**Figure 3**) [13]. It is always important to include adjacent fingers within the field of view of the image for comparison [14]. Three-Tesla MRIs are preferred due to the high resolution and detail provided for these small anatomical regions. Standard sequences used to evaluate for hand tendinous or ligamentous injury are: coronal PD, axial T1, coronal T1, sagittal T1, axial T2 and sagittal T2W sequences. When evaluating the wrist, the wrist should be at the center of the scanner with dedicated surface coils as well. Coronal images should be oriented between the radial and styloid ulnar processes and sagittal images prescribed 90° to coronals. The axial images should include approximately 2–3 cm proximal to the radiocarpal joint and at least 1 cm distal to the carpometacarpal joints [13].

4.2 Ultrasound examination technique

US of the wrist and hand are usually tailored to an area of interest, according to patient symptoms. The wrist is separated into a dorsal and ventral compartment. The hand is placed in prone position and a transverse sweep allows evaluation of the 6-extensor compartments. The hand is later supinated, allowing evaluation of the carpal tunnel and Guyon's canal.

4.3 De Quervain tenosynovitis

It is the second most common stenosing synovitis, presenting with pain and swelling at the styloid process region when moving the thumb or wrist. Anatomically, the abductor pollicis longus (APL) and extensor pollicis brevis (EPB) tendons are held within a fibro-osseous sheath called the extensor retinaculum. Repetitive trauma results in thickening of the tendons and retinaculum resulting in inflammation and edema. In some cases, a septum has been found between both tendons, thought to worsen symptoms.

On US, the APL and EPB tendons are thickened at the level of the radial styloid with increased fluid within the first extensor compartment. A halo sign has been described, secondary to peritendinous subcutaneous edema. Doppler imaging should show increased vascularity secondary to hyperemia and inflammation [15].

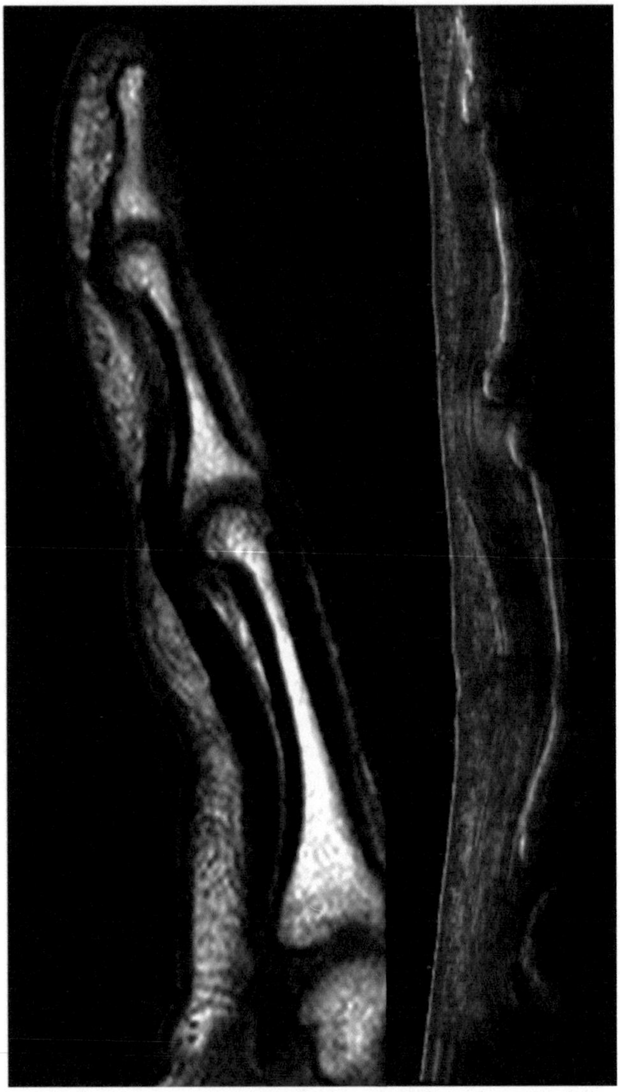

Figure 3.
On the left side, we have a T2WI showing the tendon as a hypointense structure; while on the right side we see a composite US image of the flexor tendon of the finger with some areas of anisotropy.

On MRI, tenosynovitis is seen as increased signal intensity on T2WI and low to intermediate signal on T1WI of the tendon sheath. The retinaculum will also appear thickened with increased T2 signal intensity. When the tendon is thickened, mostly seen at the radial styloid at its medial aspect, with increased T1 and T2 intra-tendinous signal and a striated tendinous signal, tendinosis is said to be present. These may also be accompanied by a longitudinal tendinous tear, where linear T2 signal will be seen traversing the tendon, most common in the APL, due to fluid within the tendinous rupture.

4.4 Flexor tendon/trigger finger

Trigger finger is a stenosing tenosynovitis secondary to repetitive microtrauma. This results in inflammation and thickening of the flexor tendon and tendon sheath, causing transient locking of the digit in a flexed position.

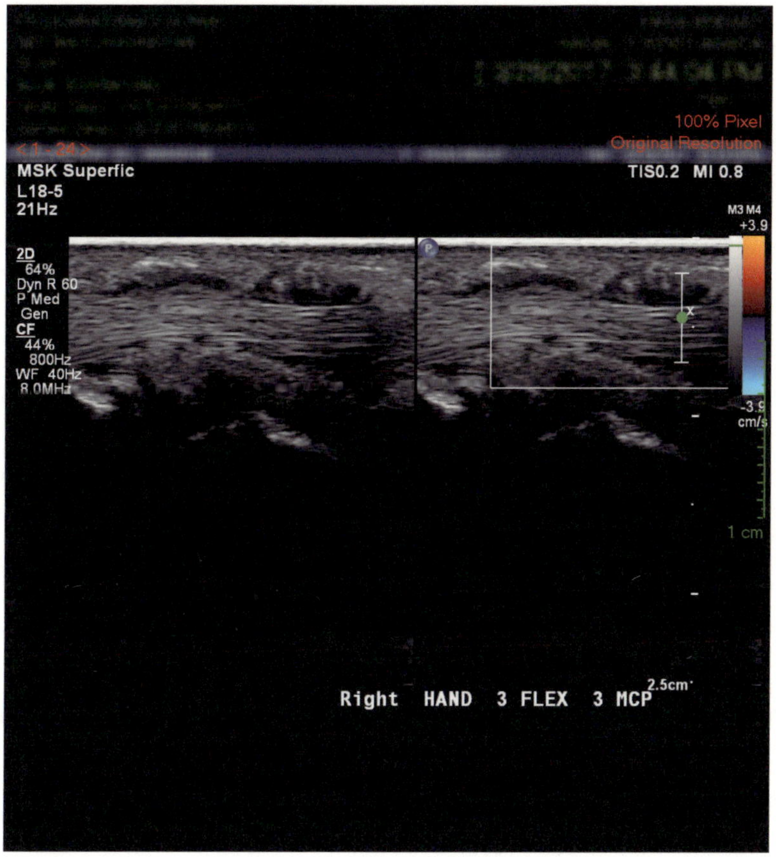

Figure 4.
Thickening of the A1 pulley in the 3rd flexor tendon of the hand consistent with clinical picture of "trigger finger".

This pathology is mainly evaluated with US instead of MRI (**Figure 4**). On US, the flexor tendon and A1 pulley will be thickened with a diameter greater than 1.1 mm. Hypoechoic fluid may also be seen around the tendon sheath, representing an effusion [16].

5. Knee

5.1 MRI protocol

The knee is positioned in a relaxed state, with about 5° of external rotation so that the anterior cruciate ligament is orthogonal to the sagittal plane of imaging. A small field of view is used, usually between 14 and 16 cm, and multiplanar imaging is obtained with coronal, sagittal, and axial images [17]. Sequences of a knee MRI should include any combination of fluid-sensitive sequences with anatomic sequences. Fluid-sensitive images can be either fat-sat PDW, or T2W spin-echo images versus STIR images. Anatomic sequences may include either T1W or PDW, spin-echo images [18].

5.2 Ultrasound examination technique

Ultrasound evaluation may be completed with the patient supine, although the posterior structures are better seen in the prone position [19]. Examination may be focused over the area that is relevant to the patient's history; nonetheless a complete examination of all areas should be performed. Sonographic examination may be divided in four methods: anterior, medial, lateral, and posterior evaluation of the knee.

1. Anterior knee: Evaluated with patient in a supine position and knee slightly flexed 20–30°. The primary structures evaluated in this approach include the quadriceps tendon, patella, patellar tendon, patellar retinaculum, suprapatellar joint recess, the medial and lateral recesses, and the anterior knee bursae [20]. Evaluation begins with transducer in the sagittal plane, proximal to the patella, to evaluate the quadriceps tendon. Deep to the quadriceps tendon, the suprapatellar recess is identified. Next, the transducer is moved inferiorly in the sagittal plane to evaluate the patellar tendon. The transducer is then moved to both the medial and lateral margins of the patella in the transverse plane, to evaluate the medial and lateral patellar retinacula, and the underlying medial and lateral recesses. Finally, the knee is placed in a 90° flexed position to evaluate the femoral trochlear cartilage in the transverse plane superior to the patella.

2. Medial knee tendons: The patient remains supine and rotates hip externally for evaluation of the medial aspect of the knee. The tendinous structures that are evaluated in this region are the pes anserine tendons [19].

3. Lateral knee: The patient is in supine position, with internal rotation of the hip, and knee slightly flexed. The key structures that are examined include the iliotibial band, lateral collateral ligament (LCL), biceps femoris tendon, popliteus, common peroneal nerve, and body and anterior horn of the lateral meniscus [19]. The transducer may be initially placed over the long axis of the patellar tendon, and then moved laterally to identify the iliotibial band. Next, the transducer is moved laterally to the coronal plane over the lateral femoral condyle to identify the groove for the popliteal tendon, an important bone landmark. Using this groove as a landmark, the proximal end of the transducer is stabilized on the femur, and the distal aspect is rotated posterior to visualize the fibular head. At this site, LCL is identified. After the transducer is moved along the LCL to its fibular attachment, the distal end of the transducer is anchored to the fibular head while the proximal aspect is rotated posteriorly in the coronal plane to visualize the biceps femoris tendon. As the transducer is moved posteriorly from the coronal plane view, the common peroneal nerve can be identified. Upon return to the popliteal groove, the distal popliteal tendon may be followed.

4. Posterior knee: The posterior aspect of the knee is evaluated with the patient in prone position and extended knee. The structures that may be identified are the posterior horns of the menisci, posterior cruciate ligament, the popliteal neurovascular bundle, and the presence of a Baker cyst [20]. The transducer is placed in the transverse plane of the mid-calf to identify the deep soleus and medial and lateral heads of the gastrocnemius muscles. The medial head of the gastrocnemius is followed proximally until the semimembranosus tendon is identified medially. If a Baker cyst is present, it will be visualized between these two structures.

5.3 Patellar tendon

The patellar tendon is part of the extensor mechanism of the knee, which originates at the patellar apex and inserts at the tibial tuberosity. It is located anteriorly to Hoffa's fat pad, and is usually about half of the thickness of the quadriceps

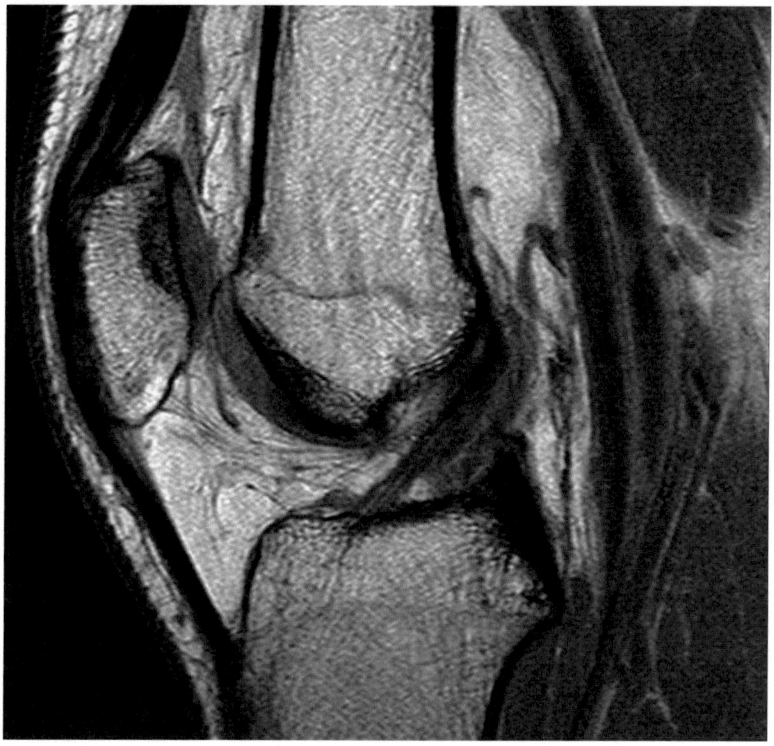

Figure 5.
Quadriceps and patellar tendons showing the dark signal qualities on PDWI.

tendon (approximately 0.5 cm), as seen on sagittal MRI with low homogeneous signal in all sequences (**Figure 5**) [18]. When visualizing it with ultrasound, the patellar tendon should normally exhibit an echogenic, fibrillar appearance. Deep to the tendon, Hoffa's fat pad appears hyperechoic or isoechoic to muscle. The region around the distal patellar tendon is also evaluated for infrapatellar bursal fluid.

Focal patellar tendinosis of the proximal deep insertional fibers is termed jumper's knee in adults, usually presenting as pain in the inferior patellar region. It is often visualized on MRI as thickening of the proximal patellar tendon with increased signal on T2W images [21]. A similar finding in children (often associated with cerebral palsy) is known as Sinding-Larsen-Johansson disease. A complete rupture of the tendon is usually easily identified, due to the secondary finding of a patella alta.

Ultrasound can be very useful in the evaluation of tendinosis and partial tears. Tendinosis will appear as focal or diffuse hypoechogenicity and thickening of the tendon. Partial-thickness tear may reveal similar findings with possible anechoic interstitial clefts. Marked hyperemia from neovascularity may also be identified with color Doppler imaging [22]. Full-thickness tears are seen as complete tendon fiber discontinuity and refraction shadowing at the retracted torn tendon stumps [20].

6. Hip joint

6.1 MRI protocol

MRI evaluation of the hip is performed while the patient is in the supine position. Coronal, axial, sagittal and axial oblique planes are obtained for

MR sequences of the entire pelvis and unilateral symptomatic hip. Imaging sequences of the hip are performed with bilateral legs placed in 15° of internal rotation. Hip MRI generally includes the following imaging sequences: non-fat saturated T1WI, fat saturated T2WI, PD fat saturated images and inversion recovery (IR) images [23].

6.2 Ultrasound examination technique

Examination can be divided in the following approaches [24, 25]:

1. Anterior hip: Patient lies supine with hip in neutral position. Examination starts with the anterior synovial recess, for which the transducer is placed over the femoral head in the oblique longitudinal plane. Examination continues with identifying the anterior glenoid labrum, located cranially in this plane, and the iliofemoral ligament that lies superficially in relation to the labrum. Next, the transducer is placed at the interphase between the femoral head and the joint space to examine the iliopsoas muscle and tendon. The neurovascular bundle and the iliopectineal eminence are used as anatomical landmarks to identify these structures. Lateral to the neurovascular bundle, the iliopsoas muscle is visualized. The iliopsoas tendon lies deep within the bellies of the muscle and on top of the iliopectineal eminence. The adjacent bursa is identified if there is a pathologic process present.

2. Medial hip: Patient remains in the supine position, now with abduction and external rotation of the hip and flexion of the knee. Examination starts in the long axis plane to scan over the insertion of the iliopsoas tendon at the lesser trochanter of the femur. Next, the adductor muscles are evaluated in the axial plane. The muscles of the medial hip compartment are divided in three layers. The adductor longus is located at the lateral aspect of the superficial muscular layer, while the gracilis is located at the medial aspect. The adductor brevis makes up the intermediate muscular layer and the adductor magnus makes up the deep muscular layer. Scan continues in the long axis plane with the transducer moved along the abductor muscles to identify the abductor longus tendon, using the pubic bone as reference landmark. The adductor longus tendon insertion is identified as a hypoechoic triangular structure. Lastly, the transducer is placed over the pubis in the transverse plane, from which oblique longitudinal plane is achieved to evaluate the tendon complex formed by the transversus abdominis and internal oblique muscles.

3. Lateral hip: The patient is moved to the lateral decubitus position, lying on opposite hip of interest. With this examination, the following structures are evaluated: abductor muscles, gluteus medius, gluteus minimus, and tensor fascia lata. To begin, the transducer is placed over the greater trochanter. Scanning is performed in the transverse and longitudinal planes. The gluteus medius, seen as a curvilinear fibrillar band, lies superficial to the gluteus minimus. The tensor fascia lata serves as an anatomical landmark to identify the gluteus muscles, which is visualized as a superficial hyperechoic band in the coronal plane.

4. Posterior hip: Evaluated with the patient in the prone position. Important structures to evaluate include: the hamstring muscles and the sciatic nerve. Examination starts in the transverse plane with the transducer positioned at the ischial tuberosity to identify the hamstring tendon complex, where no distinction can be made between each individual tendon. The sciatic nerve is a lateral flattened structure with fascicular echotexture. As the transducer is

DOI: http://dx.doi.org/10.5772/intechopen.84521

moved caudally, distinction between individual hamstring tendons is achieved with the semimembranosus tendon located deep and medially in relation to the conjoined tendon complex of the biceps femoris and semimembranosus. Sonographic appearance of this conjoined tendon complex is a hyperechoic line that separates the laterally located biceps femoris muscular belly from the medially located semimembranosus muscular belly.

6.3 Gluteal tendons

The gluteus medius and gluteus minimus tendons are part of the lateral compartment of the hip. The gluteus medius tendon inserts at the lateral and superoposterior facets of the greater trochanter of the femur, while the gluteus minimus tendon inserts at the anterior aspect of the greater trochanter. Ultrasound shows the gluteus medius tendon as a hyperechoic structure arising from a fan shaped hypoechoic structure that represents the gluteus medius muscle. The gluteus medius and minimus muscles are separated by an echogenic layer of fascia and adipose tissue [26].

The gluteus medius and minimus are the most commonly affected tendons of the hip abductor group that cause greater trochanteric pain syndrome [28]. Gluteus tendon abnormalities may be due to acute injury or chronic wear and tear of the hip joint. Therefore, it typically affects women in the middle and elderly age groups.

MRI is the gold standard imaging modality for the identification of gluteus tendon tears (**Figure 6**). Axial and coronal T2W fat saturated images of the hip and coronal T1WI of the pelvis are recommended when abductor tendon pathology is suspected [29]. MRI diagnostic criteria for tendon tears include discontinuity of the tendon, elongation of the gluteus medius 2 cm or greater and T2 hyperintensities superior to the level of the greater trochanter of the femur [27]. Additional MR findings, although nonspecific, may include atrophy of the adipose tissue, changes of the adjacent bone structures and fluid collection within the bursa.

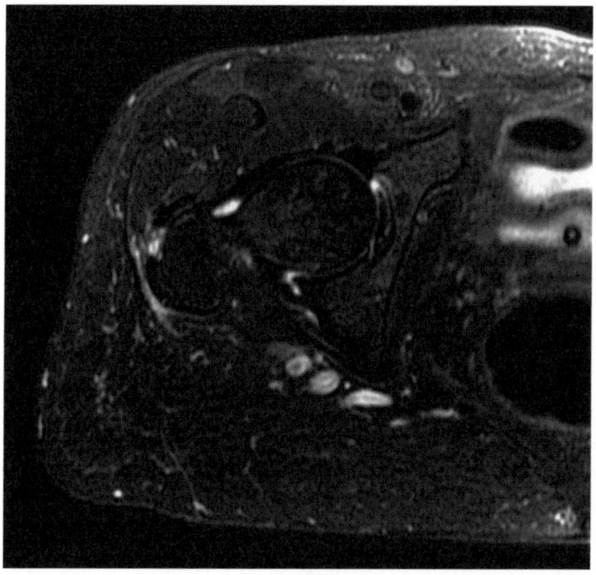

Figure 6.
T2 fat saturated axial image shows a full thickness tear of the gluteus medius tendon as a fluid-filled defect along the greater trochanter.

6.4 Iliopsoas tendon

The psoas major and iliacus muscles form the iliopsoas tendon complex. The psoas major originates at the transverse processes of L1-L5 vertebrae; while the iliacus muscle has various origins, including the superior two thirds of the iliac fossa, the anterior sacroiliac ligaments, and the anterior sacral ala. This tendon complex inserts at the lesser trochanter of the femur.

The iliopsoas tendon has an echogenic sonographic imaging appearance, with anterior extension in relation to the anterior-superior acetabular labrum [28]. On MRI sequences, the iliopsoas tendon complex is characteristically identified as two parallel homogeneously hypointense structures, separated by a hyperintense region that represents adipose tissue of the fascia [28].

Snapping iliopsoas tendon is characterized by an audible or palpable painful snap with movement of the hip. Repetitive movements of the hip serve as predisposition to develop a snapping tendon, such as those performed by young athletes in different sports, with ballet dancers being the most commonly affected [30]. Iliopsoas tendon as the source of a snapping hip is classified as an internal cause of the broader term snapping hip syndrome. It may get trapped during movement due to a prominent iliopectineal eminence, an insertion site osseous projection or the anterior inferior aspect of the iliac spine [30].

Dynamic evaluation of the hip joint with US and MRI allows the identification of the source of snapping iliopsoas tendon. Sonographic evaluation is performed with a high-frequency transducer (linear 5–12 MHz) placed in the transverse oblique plane, above the hip joint and parallel to the pubis [28]. The patient is in the supine position, with initial static evaluation performed following the iliopsoas tendon until reaching its insertion at the lesser trochanter. Dynamic evaluation is performed while the ipsilateral leg is moved from the "frog leg" position (extension, adduction, and internal rotation) to the neutral position (flexion, abduction, and external rotation). The position of the iliopsoas tendon can be traced along the anterior compartment of the hip as the leg is moved from the aforementioned positions and snapping occurs. Regarding MRI examination for this particular pathology, fast GRE sequence allows dynamic evaluation of the iliopsoas tendon during movement [28]; change from "frog leg" to neutral position is also performed during this MRI sequence.

6.5 Hamstring tendons

The hamstring tendon complex is located at the posterior compartment of the hip, formed by three muscles groups: biceps femoris, semimembranosus and semitendinosus. The biceps femoris is composed of a short and a long head. Origin of the short head is at the lateral linea aspera of the posterior femur, the lateral supracondylar line and the intermuscular septum [31]. The long head shares origin with the semitendinosus tendon at the inferomedial facet of the ischial tuberosity to form a conjoint tendon. Distal biceps femoris tendon inserts at the lateral fibular head and lateral condyle of proximal tibia, while the semitendinosus inserts at the anteromedial aspect of the tibia, sharing insertion with the gracilis and sartorius muscles to form the pes anserinus tendons [32]. The semimembranosus tendon originates at the superolateral ischial tuberosity and has several insertion sites through tendinous arms [31]. The anterior, direct, and inferior arms insert at the medial condyle of the tibia. The capsular arm inserts at the posterior oblique ligament. There is also insertion into the posterior joint capsule and arcuate ligament through the oblique popliteal ligament.

On MRI sequences at the level of the ischial tuberosity, the hamstring tendons are identified as well-defined round areas of low signal intensity, where the conjoint tendon is posteromedial to the semimembranosus tendon [31].

The hamstrings are the most commonly injured muscle group in athletes, with tendon avulsion as the most severe injury diagnosed with medical imaging, requiring prompt surgical management. Avulsion injuries are defined as complete tear of the tendon from its osseous insertion site and typically affect the hamstrings proximally, particularly the conjoint tendon. This type of injury can include pulling of a bone piece by the torn tendon, which most commonly occurs in children due to presence of growth plates. MRI is the gold standard for examination of suspected hamstring tendon avulsion. Evaluation approach involves identifying the affected tendon and determining whether a partial or full thickness tear occurred. In the case of full thickness tears, distal tendon retraction, degree of underlying tendinopathy, and proximity of the tear to the sciatic nerve must be included in the evaluation approach [32].

7. Ankle

7.1 MRI protocol

The ankle tendons are visualized as low signal intensity structures in all MR sequences. The T1WI sequences are used to evaluate the anatomy and the T2W sequences are used to assess abnormal increase in fluid, usually related to tendon pathology [33]. Axial images are used to evaluate morphologic features of the tendons and synovial sheath distention, longitudinal splits, fluid within the tendon sheath, and adjacent soft tissue abnormalities, if any. For evaluating the Achilles tendon, sagittal images prove most useful. Sagittal images also assess the proximal-to-distal extent of tendon pathologies. Oblique coronal or short axis images at the level of the mid- and forefoot are best for assessment of the tendons distal to the ankle [33, 34].

When the normal tendons form an angle of approximately 55° with the main magnetic vector, it produces increased signal intensity within the tendons. This phenomenon is called the magic angle, more commonly in sequences with echo times less than 20 msec (T1WI, PD or GRE). This effect is particularly common with ankle tendons because of their curvatures around the ankle joint [33, 34].

For general purposes, an ankle MRI should include at least the following: axial T1WI or PD sequences and fat-suppressed T2WI, coronal T1WI or fat-suppressed T2WI and IR sagittal images [34].

7.2 Ultrasound examination technique

1. Peroneal tendon: Evaluated with the patient in the supine position with the knee semi-flexed and the ankle in internal rotation. For evaluation of the plantar aspect of the peroneus longus tendon, the patient should be in the prone position [35]. Both peroneal tendons are examined with linear transducers in their short and long planes. The transducer is placed behind the lateral malleolus over the tendons to examine their short-axis first. The transducer should be tilted along the way, to maintain the perpendicular position of the US beam. The tendons should be evaluated upwards for approximately 5 cm and downwards into the inframalleolar region [34, 36, 37].

2. Posterior tibial tendon: Evaluated with the patient in a seated position with internal rotation of the plantar surface of the foot. If this position cannot be

achieved, the patient may lie supine with the foot slightly laterally rotated. Placing the transducer in short-axis/transverse position behind the medial malleolus, evaluate the posterior tibial tendon following it from its myotendinous junction into its insertion [36].

3. Achilles tendon: Evaluated with the patient in prone position and the foot hanging from the examination table. With the transducer, follow the tendon from its myotendinous junction downward to its calcaneal insertion in both the short and long planes. The size of the tendon should only be obtained in the transverse plane [36].

7.3 Peroneal tendon

The peroneal tendons are the third most commonly injured tendons of the ankle. Acute and chronic tears occur in young, athletic patients due to overuse or in older patients with multifactorial degenerative wear and tear. Due to their course and location, calcaneal fractures predispose to partial tears, entrapment and dislocation of the peroneal tendons. Tendinopathy more commonly affects the peroneus brevis tendon. Split peroneus brevis syndrome represents a longitudinal tear of the tendon and the term arises from the fact that the peroneus brevis tendon is usually located anteriorly, is embedded between the peroneus longus and fibula [33, 34, 38].

7.4 Tibialis posterior tendon

The tibialis posterior tendon is the second most commonly injured of the ankle tendons [37]. It should never have more than twice the cross-sectional area of the flexor digitorum longus tendon. Posterior tibial tendinopathy occurs because of delayed stretching of the tendon due to chronic micro-tears, and usually occurs in older women with progressively painful flat-foot. Systemic diseases like rheumatoid arthritis and diabetes predispose this condition, as for other tendinopathies. On MR and ultrasound imaging it will appear as tendon thickening with loss of normal echogenicity and tendon sheath fluid, with increased T2 signal intensity. Imaging pitfalls include: normal tendon widening at its insertion onto the navicular bone; fluid within the tendon sheath, mimicking enlargement on T1W sequences; and magic angle phenomenon [33, 37, 39]. Due to its course and insertion, abnormalities of the navicular bone may predispose to tibialis posterior tendinopathy; like the type II accessory navicular bone or *os naviculare,* which is typically large and closely positioned at the medial pole of the navicular bone by a synchondrosis, rendering insertion of the posterior tibial tendon only on this ossicle and not extending into the cuneiforms and metatarsals [36].

7.5 Achilles tendon

The Achilles tendon is the most commonly injured tendon of the ankle (**Figure** 7) [37]. It is usually hypointense on all MR sequences, although due to its fascicular anatomy, a single line may be visible (not on T2WI), mimicking an interstitial tear. Punctate foci of increased signal intensity may be noted in axial images of the distal Achilles tendon, which simply are interfascicular membranes. Normal average thickness is 6–8 mm, which may increase in male, tall and elderly patients. On axial images its margins are concave for the majority of its course, being more convex proximally to and at the soleus insertion. Normally, there should be subcutaneous fat between the Achilles tendon and the skin. Branches of the posterior tibial artery supply the Achilles tendon, but blood supply diminishes at approximately 2–6 cm proximal to its insertion site, making this region of decreased vascularity particularly susceptible to ruptures [35, 37].

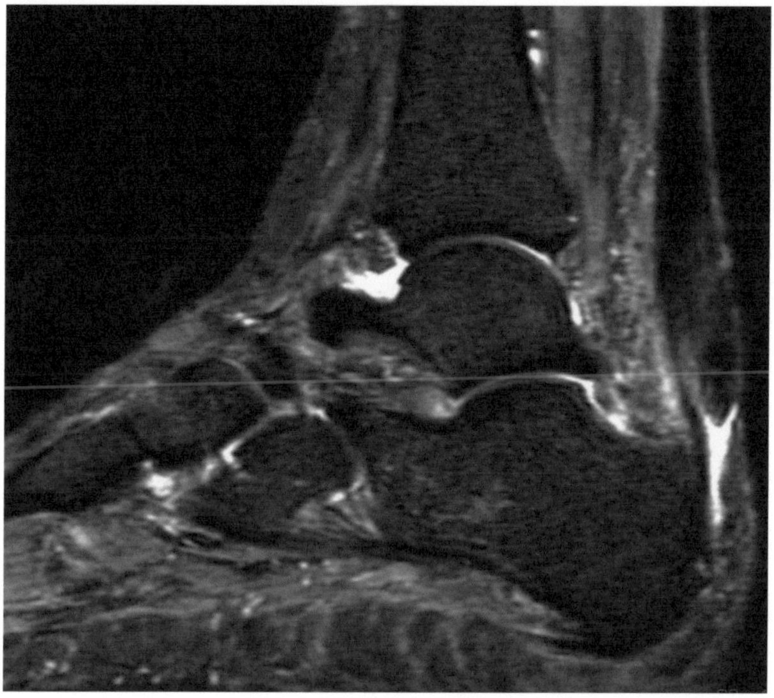

Figure 7.
T2 fat saturated sagittal image shows disruption of the distal Achilles tendon with a fluid-filled gap.

Achilles tendinosis is common in runners and jumpers. In Achilles tendinosis and peritendinosis, the tendon may enlarge. Acute Achilles ruptures more commonly occur in patients with chronic tendinopathy; runners, middle-aged women who engage in sporadic exercise or patients with systemic diseases or chronic steroid use, resulting in a weakened tendon. Most common site of Achilles tendon rupture is 2–6 cm proximal to its insertion site, avascular zone, as detailed above. Acute ruptures show a tendon gap with intermediate signal intensity on T1WI and increased signal intensity on T2WI, consistent with edema and hemorrhage. In chronic ruptures the gap is replaced by fat and scar tissue [33, 35, 37]. An accessory soleus muscle may be mistaken with a thickened Achilles tendon, which differ by their signal intensity on MRI. Achilles tendon thickening may occur after surgical procedures. There is thickening also with of xanthomas (familial hyperlipidemia) that appear as marked tendon enlargement with heterogeneous signal masses and linear areas of low signal intensity. Haglund's disease results most commonly from ill-fitting shoes that compress the distal Achilles tendon, leading to peritendinous edema, retrocalcaneal bursitis and tendon thickening [37].

8. Conclusion

MRI and US are useful imaging modalities that allow anatomic evaluation of tendons as well as identification of tendon pathology.

Acknowledgements

Thanks to the University of Puerto Rico School of Medicine Diagnostic Radiology Department for allowing us to use images from its teaching file.

Conflict of interest

None.

Acronyms and abbreviations

MRI	magnetic resonance imaging
US	ultrasound
T1WI	T1-weighted image
FSE	fast spin echo
T2WI	T2-weighted image
PD	proton density
GRE	gradient echo
APL	abductor pollicis longus
EPB	extensor pollicis brevis
LCL	lateral collateral ligament
IR	inversion recovery

Author details

Stephanie C. Torres-Ayala[1*], Alvaro Bravo-Martínez[1], Amanda P. Marrero-González[1], Luis R. Rodriguez-Ortíz[1] and Jorge A. Vidal[2,3]

1 University of Puerto Rico School of Medicine Diagnostic Radiology, San Juan, Puerto Rico, USA

2 University of Puerto Rico School of Medicine, Musculoskeletal Radiology Section, San Juan, Puerto Rico, USA

3 University of Alabama at Birmingham, Emergency Radiology Section, Birmingham, Alabama, USA

*Address all correspondence to: stephanie.torres19@upr.edu

References

[1] Chen Q, Miller TT, Padron M, Beltran J. Normal shoulder. In: Musculoskeletal Imaging. 2nd ed. Philadelphia, PA: Elsevier; 2015. pp. 70-86

[2] Singh JP. Shoulder ultrasound: What you need to know. Indian Journal of Radiology and Imaging. 2012;**22**(4):284-292. DOI: 10.4103/0971-3026.111481

[3] Jacobson JA. Shoulder US: Anatomy, technique, and scanning pitfalls. Radiology. 2011;**260**(1):6-16. DOI: 10.1148/radiol.11101082

[4] Crass JR, Craig EV, Feinberg SB. The hyperextended internal rotation view in rotator cuff ultrasonography. Journal of Clinical Ultrasound. 1987;**15**(6):416-420

[5] Ferry M, Finlay K, Popowich T, Stamp G, Schuringa P, Friedman L. Sonography of full-thickness supraspinatus tears: Comparison of patient positioning technique with surgical correlation. American Journal of Roentgenology. 2005;**184**(1):180-184. DOI: 10.2298/SARH0912647S

[6] Arend CF, Silva TR. Comparison between exclusively long-axis and multiple-axis sonographic protocols for screening of rotator cuff lesions in symptomatic shoulders. Journal of Ultrasound in Medicine. 2010;**29**:1725-1732

[7] Helms CA, Major NM, Anderson MW, Kaplan PA, Dussault R. Shoulder. In: Musculoskeletal MRI. 2nd ed. Philadelphia, PA: Elsevier; 2009. pp. 177-223

[8] Jacobson JA. Shoulder ultrasound. In: Fundamentals of Musculoskeletal Ultrasound. 3rd ed. Philadelphia, PA: Elsevier; 2018. pp. 55-126

[9] Skendzel JG, Jacobson JA, Carpenter JE, et al. Long head of biceps brachii tendon evaluation: Accuracy of preoperative ultrasound. AJR. American Journal of Roentgenology. 2011;**197**:942-948. DOI: 10.2214/AJR.10.5012

[10] Sampath SC, Sampath SC, Bredella MA, et al. Magnetic resonance imaging of the elbow. Sports Health. 2013;**5**(1):34-39. DOI: 10.1177/1941738112467941

[11] Konin GP, Nazarian LN, Walz DM. US of the elbow: Indications, technique, normal anatomy, and pathologic conditions. Radiographics. 2013;**33**(4):125-127. DOI: 10.1148/rg.334125059

[12] Bucknor MD, Stevens KJ, Steinbach LS. Elbow imaging in sports: Sports imaging series. Radiology. 2016;**279**(3):827-837. DOI: 10.1148/radiol.2016151256

[13] Kassarjian A, Benjamin L, Afonso D, et al. Guidelines for MR Imaging Sports Injuries. European Society of Skeletal Radiology Sub-committee [Internet]. 2016. Available from: https://essr.org/content-essr/uploads/2016/10/ESSR-MRI-Protocols-Thumb.pdf

[14] Gupta P, Lenchik L, Wuertzer SD, et al. High-resolution 3-T MRI of the fingers: Review of anatomy and common tendon and ligament injuries. American Journal of Roentgenology. 2015;**204**(3):W314-W323. DOI: 10.2214/AJR.14.12776

[15] Glajchen N, Schweitzer M. MRI features in de Quervain's tenosynovitis of the wrist. Skeletal Radiology. 1996;**25**(1):63-65. DOI: 10.1007/s002560050033

[16] Guerini H, Pessi E, Theumann N, et al. Sonographic appearance of trigger fingers. Journal of Ultrasound in Medicine. 2008;**27**(10):1407-1413. DOI: 0278-4297/08/$3.50

[17] Helms CA, Major NM, Anderson MW, Kaplan PA, Dussault R. Knee. In: Musculoskeletal MRI. 2nd ed. Philadelphia, PA: Elsevier; 2009. pp. 353-383

[18] De Maeseneer MO, Shahabpour M. Normal knee. In: Musculoskeletal Imaging. 2nd ed. Philadelphia, PA: Elsevier; 2015. pp. 324-332

[19] Alves TI, Girish G, Kalume brigido M, Jacobson JA. US of the knee: Scanning techniques, pitfalls, and pathologic conditions. Radiographics. 2016;**36**(6):1759-1775. DOI: 10.1148/rg.2016160019

[20] Jacobson JA. Knee ultrasound. In: Fundamentals of Musculoskeletal Ultrasound. 3rd ed. Philadelphia, PA: Elsevier; 2018. pp. 284-327

[21] Yus JS, Popp JE, Kaeding CC, Lucas J. Correlation of MR imaging and pathologic findings in athletes undergoing surgery for chronic patellar tendinitis. American Journal of Roentgenology. 1995;**165**:115-118. DOI: 10.2214/ajr.165.1.7785569

[22] Khan KM, Bonar F, Desomnd PM, et al. Patellar tendinosis (jumper's knee): Findings at histopathologic examination, US, and MR imaging-Victorian Institute of Sport Tendon Study Group. Radiology. 1996;**200**(3):821-827. DOI: 10.1148/radiology.200.3.8.8756939

[23] Kassarjian A, Fritz B, Afonso PD, Alcalá-Galiano A, Ereño MJ, Grainger A, et al. Guidelines for MR Imaging of the Hip Region [Internet]. 2016. Available from: essr.org/content-essr/uploads/2016/10/ESSR_Sports_guidelines.pdf [Accessed: November 19, 2018]

[24] Beggs I, Bianchi S, Bueno A, Cohen M, Court-Payen M, Grainger A, et al. Musculoskeletal Ultrasound Technical Guidelines IV. Hip [Internet]. 2016. Available from: https://essr.org/content-essr/uploads/2016/10/hip.pdf [Accessed: November 19, 2018]

[25] Lin YT, Wang TG. Ultrasonographic examination of the adult hip. Journal of Medical Ultrasound. 2012;**20**(4):201-209. DOI: 10.1016/j.jmu.2012.10.009

[26] Connell DA, Bass C, Sykes CA, Young D, Edwards E. Sonographic evaluation of gluteus medius and minimus tendinopathy. European Radiology. 2003;**13**(6):1339-1347. DOI: 10.1007/s00330-002-1740-4

[27] Cvitanic O, Henzie G, Skezas N, Lyons J, Minter J. MRI diagnosis of tears of the hip abductor tendons (gluteus medius and gluteus minimus). American Journal of Roentgenology. 2004;**182**(1):137-143. DOI: 10.2214/ajr.182.1.1820137

[28] Bancroft LW, Blankenbaker DG. Imaging of the tendons about the pelvis. American Journal of Roentgenology. 2010;**195**(3):605-617. DOI: 10.2214/ajr.10.4682

[29] Hartigan DE, Perets I, Walsh JP, Domb BG. Imaging of abductor tears: Stepwise technique for accurate diagnosis. Arthroscopy Techniques. 2017;**6**(5):e1523-e1527. DOI: 10.1016/j.eats.2017.06.032

[30] Piechota M, Maczuch J, Skupinski J, Kulawska-Sysio K, Wawrzynek W. Internal snapping hip syndrome in dynamic ultrasonography. Journal of Ultrasonography. 2016;**16**(66):296-303. DOI: 10.15557/JoU.2016.0030

[31] Koulouris G, Connell D. Hamstring muscle complex: An imaging review. Radiographics. 2005;**25**(3):571-586. DOI: 10.1148/rg.253045711

[32] Rubin DA. Imaging diagnosis and prognostication of hamstring injuries. American Journal of Roentgenology. 2012;**199**(3):525-533. DOI: 10.2214/ajr.12.8784

[33] Rosenberg ZS, Beltran J, Bencardino JT. From the RSNA refresher courses MR imaging of the ankle and foot. Radiographics. 2000;**20**:153-179. DOI: 10.1148/radiographics.20.suppl_1.g00oc26s153

[34] Wang X-T, Rosenberg ZS, Mechlin MB, Schweitzer ME. Normal variants and diseases of the peroneal tendons and superior peroneal retinaculum: MR imaging features. Radiographics. 2005;**25**:587-602. DOI: 10.1148/rg.253045123

[35] Schweitzer ME, Karasick D. MR imaging of disorders of the Achilles tendon. American Journal of Roentgenology. 2000;**175**:613-625

[36] Beggs I, Bianchi S, Bueno A, Cohen M, Court-Payen M, Grainger A, et al. Musculoskeletal ultrasound: technical guidelines. Insights into Imaging. 2010;**1**:99-141. DOI: 10.1007/s13244-010-0032-9

[37] Manaster BJ, May DA, Disler DG. Musculoskeletal Imaging: The Requisites. 4th ed. Saunders: Elsevier; 2013. pp. 215-234. DOI: 978-0-323-08177-1.ch14

[38] Taljanovic MS, Alcala JN, Gimber LH, Rieke JD, Chilvers MM, Latt LD. High-resolution US and MR imaging of peroneal tendon injuries. Radiographics. 2015;**35**:179-199. DOI: 10.1148/rg.351130062

[39] Schweitzer ME, Karasick D. MR imaging of disorders of the posterior tibialis tendon. American Journal of Roentgenology. 2000;**175**:627-635. DOI: 10.2214/ajr.175.3.1750627

Chapter 4

Exercise and Tendon Remodeling Mechanism

Rita de Cassia Marqueti, Ivo Vieira de Sousa Neto, Fabricio Reichert Barin and Gracielle Vieira Ramos

Abstract

Tendons connect muscles to bones and transmit the force exerted by the corresponding muscle to the skeleton and, therefore, are key components for locomotion. They are responsive to mechanical factors, which are essential for cellular functioning, tendon development, homeostasis, and repairing. Mechanical signals are transduced via molecular signaling pathways which trigger tendon adaptive responses. Previous data have already shown that exercise training promotes physiological adaptive responses, such as morphological properties and biomechanical and biochemical adaptations.

Keywords: tendon, exercise, extracellular matrix, mechanical loading

1. Tendon macro- and microscopic overview

Anatomically, tendons are located in muscle origin/insertion and, further, in tendon intersections within the muscles. These complexly arranged tissues have great importance in movement generation, having as main actions the transfer of tension produced by the muscles to the subsequent unit, thus determining the degree of articular movement produced [1, 2]. Moreover, tendons act as a mechanism of energy storage, improving movement economy and power amplification for activities as jumping and acceleration, based on their spring-like properties [1, 3]. More recently, studies have shown that a stretch applied to an active muscle-tendon unit (MTU) can be taken up solely by tendon stretch, with little or no muscle fascicle lengthening, acting like a mechanical buffer to protect muscle fascicles, attenuation damage associated with active lengthening [4]. In addition, the energy absorbed by the tendons during movement can be used to maximize the power generated, as it can also be dissipated as heat, increasing energy release time to the muscles, thus decreasing the peak force experienced by the MTU [3–5]. All these functional aspects can promote more efficient movements, consuming less energy and preventing muscle damage [1, 4, 5] (**Figure 1**).

The tendinous actions require an unrestricted slip by the adjacent tissue. For this reason, synovial sheaths form a closed system around many tendons to provide lubrication and to cushion the tendon during its action. Some tendons that do not have this system find in their periphery a loose peritendinous adipose and vascularized layer, which allows the free excursion of the tendon [6]. When healthy, they present white color, due to the presence of hypovascularized zones and fibroelastic texture, characterizing this tissue resistance to mechanical stresses [7]. In relation to

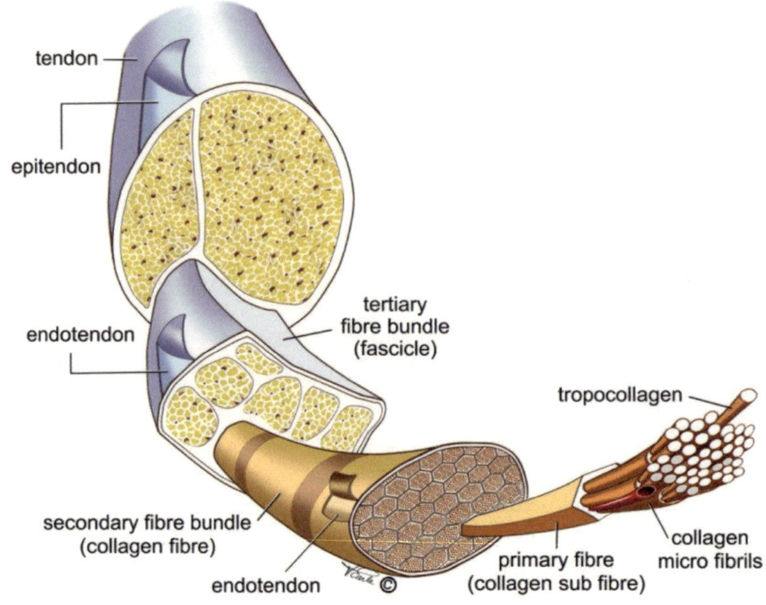

Figure 1.
Tendon hierarchical structure (adapted from Kannus, [8]).

its composition, because it is a connective tissue, the division happens between cells and the extracellular matrix (ECM) [7]. The ECM tendon is formed by approximately 70% of water, with great part associated with proteoglycans, glycosaminoglycans (GAGs), adhesion glycoproteins, non-collagenous proteins, and fibrous proteins (collagen and elastin) [1]. The water and GAG presence in the tendons is extremely important for maintaining the spacing among the collagen fibers, facilitating their slippage [9]. The highly organized collagen bundles represent 60–95% of the dry weight of the tendon, from which 90% are type I collagen and the remainder part is divided into types III (0–10%), IV (2%), V, and VI, respectively [2, 9].

The collagen formation occurs through the three spiral and helical peptide chains. Structural formation occurs intracellularly, with subsequent secretion into the ECM, where it is converted into collagen. The fibrillar structure is stabilized by various post-translational modifications, allowing the formation of intermolecular and interfibrillar cross-links. These cross-links' large amount makes this structure highly resistant to stresses, shear forces, and even compression [10]. There are two main types of cross-links—enzymatic cross-links, confined in the terminal domains, and advanced glycation end-products (AGEs)—which may be formed at various sites along the collagen length [11]. It has been shown that enzymatic cross-links are essential in the formation and functioning of collagen fibrils, stabilizing the structure, while AGEs accumulate with age and diabetes and may impair the normal function of fibrils, leading to decreased viscoelasticity of the tendon [12, 13].

Since tendons are attached to two totally distinctive structures in morphological and functional terms (muscles and bones), two types of tendon junctions are found and differ in relation to the number of cross-links. The myotendinous junction (MJT) is more compliant, thus having smaller amount of cross-links, so that the perfect junction can occur [14, 15]. This decrease occurs due to increased shear forces, mainly from muscle activity. On the other hand, the osteotendinous junction (OJT) is less compliant, with much greater amount of cross-links [16]. This regional difference is also identified by biomechanical studies, which show greater stress in the tendon distal region, when compared to the proximal region. Thus, a study proposes that, besides that the tendon acts as a protective factor for the muscle (mechanical buffer), it seems that the tendon proximal region acts as a mechanical buffer of the tendon distal region, which undergoes greater mechanical loads and has greater stiffer [17].

Type I collagen molecules are generally heterotrimeric, composed of two α1 polypeptide chains and one α2 polypeptide chain, interlaced in a triple helix, as previously mentioned. Studies have shown that the α2 chain is the most hydrophobic, thus playing a stabilizing role for this molecule [18]. The type III collagen (second most commonly found type) is located mainly in the epitendon and endotendon. Authors have shown that their proportion changes under conditions of tissue injury. Among its functions, types I and V collagen heterotypic fibril formation and fibrillar diameter control stand out [1]. Type IV collagen is found primarily in the basement membrane, where it is considered the main structural component.

A small amount of elastin (2%) still contemplates the composition of tendon. This low proportion is directly related to the almost inelastic aspect of the tendon, its extensibility varying between 8 and 10% before reaching the point of total rupture. This aspect is extremely important, so that the function of conducting tension, performed by the tendon, is efficient [2, 9]. The tendinous arrangement is considered to be a highly organized hierarchical structure [9]. The tropocollagen molecules are organized and grouped into microfibrils and, later, into collagen fibrils. These primary beams bind to the formation of fascicles (smaller functional units of the tendon). Covering the fascicles is the endotendon, a layer of connective tissue, with nerve and vascular and lymphatic structures. By grouping the endotendons in an organizational way, we find the epitendon, the layer responsible for involving the tendon itself. Finally, the paratendon, a more vascularized layer, which can become a double layer, is filled with synovial fluid (produced by the synovial membrane) in tendons subject to friction [19, 20].

In relation to non-collagenous proteins, proteoglycans and glycoproteins are essential to ensure the binding between collagen fibers, water molecule diffusion, and collagen fibrillogenesis and to maintain the matrix structure, being essential for tissue viscoelastic capacity [19]. There are two groups of proteoglycans in the tendon, accounting for 1–5% of the dry weight of the tendon, with high hydrophilicity, attracting water molecules. The small leucine-rich proteoglycans (SLRPs) (decorins, fibromodulins, biglicans, and lumicans), which have a small core protein (40 kDa) and large proteoglycans, also called modular proteoglycans (aggrecan and versican), have a core protein of about 220–370 kDa. The proteoglycan amount in the tendon can vary, depending on the tension and compression (mechanical stimuli) that it receives [21]. Among the glycoproteins, tenascin-C acts on the stability and structuring of ECM, and its expression is greater in tissues where ECM turnover is more active. About fibronectin, it serves as a bridge between cells and ECM [22].

Regarding the cellular environment, the tenoblasts and tenocytes comprise 90–95% of the elements, together with endothelial cells and mast cells [1, 9]. The tenoblasts are spindle-shaped cells, immature, and highly metabolic and present many cytoplasmic organelles. These cells, over time, differentiate into tenocytes, more elongated cells, and are distributed in rows among the collagen fiber bundles, located mainly in the peritendinous region [23, 24]. The tenocytes are arranged in parallel rows along the force transmission axis and interact with the ECM through the integrins (adhesion cell surface receptors) that connect the intracellular cytoskeleton to the matrix, allowing the propagation of mechanical signals from the outside to the inside and from the inside to the outside (bidirectional), a process known as mechanotransduction, that is, the cellular capacity to feel and respond to external mechanical stimuli, extremely important for the maintenance of the tendinous function, besides the organization of its structure [25]. Additionally, tenocytes have extensive communication with adjacent cells through gap junctions. The communication junctions are extremely complex structures, having two hemicanals, also called connexons, with a central pore. These open connexons allow free metabolites and ion passage among the gap junctions [26, 27]. These connexons are numbered, and the most important ones, in terms of cellular communication and tendons regeneration, are those of number 32 and 43. Connexin 43 is responsible for the inhibition of the collagen syntheses within the tenocytes, as a response to mechanical loading. Connexin 32 may have a stimulatory role, but all we need to know is that they aid communication among cells within the tendon to help with regeneration and adaptation [27].

2. Biomechanical aspects inherent to tendon function

Tendons have biomechanical characteristics similar to springs. They are not able to produce mechanical energy, but are able to conserve energy, increase potency during functional activities, and absorb external forces to prevent injury in the muscle [1, 3, 4]. The parallel arrangement of its collagen fibers along the tendon long axis is directly correlated with its ability to control the tensile loads, mainly unidirectionally and making tendon highly effective in transmitting tension generated by the skeletal muscle to the bone. A good example of the opposite is the ligaments, which have an interlocking arrangement, thus having greater ability to control tension loads in different directions [1, 9].

Nevertheless, a look with microscopic scales identifies a zigzag pattern of collagen fibers, somewhat different from the perfect alignment along the tendon long axis described earlier, known as crimps [28]. Crimps form a "crimp angle" of 20° from the long axis of the tendon. When tendons are unloaded or in the low-load "toe region" of the stress-strain curve (this topic will be better described below), collagen crimps that are present, and they "disappear" when loaded to induce a tensile strain of about 4%. Thus, crimps are physiological markers of tendon tension [29, 30].

The ECM composition characterizes this tissue as viscoelastic, guaranteeing the return to its original size, after being submitted to a certain level of deformation force [31, 32]. Tendon deformation, which occurs during movement, is dependent on the applied load. With higher stress levels, the tendons deform less and become stiffer, maximizing their ability to carry mechanical loads. Otherwise, with lower tensions, tendons have greater ability to deform and thus generate greater gains and adaptations in their ability to absorb energy [33]. This feature guarantees metabolic expenditure reduction during locomotion, as well as strength and power maximization during movement, a strategy known as the stretch-shortening cycle, where there is the use of the elastic capacity of passive structures, such as the tendon, for the energy accumulation during its elongation, transformed into a more powerful movement and with less energy expenditure during muscle shortening (concentric contraction) [34, 35].

For a better understanding of the biomechanical aspects of this tissue, experimental tests are performed, trapping the muscle-tendon complex at one end and the bone at the other end [36]. The tendon is then exposed to a controlled load along its longitudinal axis, recording force, and displacement until tissue failure. The results about mechanical properties are described in four main ways in literature [37]. The tendon deformation/elongation capacity relative to its normal length is characterized as strain, while the tendon force relative to its cross-sectional area is known as stress. Changes in tendon length, relative to the forces applied on it, generate stiffness. Finally, Young's modulus, or modulus of elasticity, which describes the relationship between tendon stress and strain, represents the properties, independently of the cross-sectional area (CSA) [24, 36, 38].

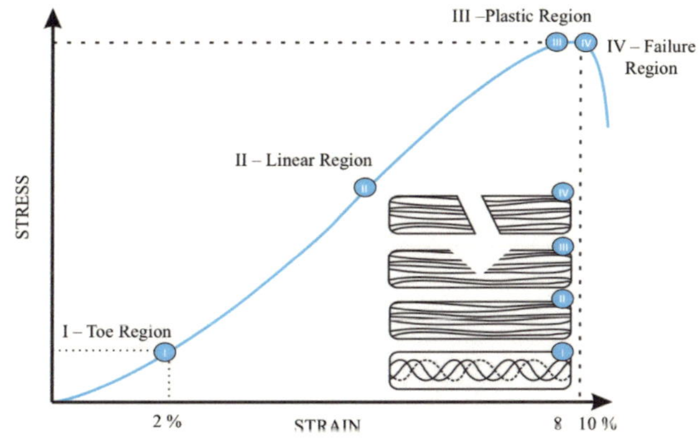

Figure 2.
Stress-strain curve and its four distinct regions: (I) toe region, (II) linear region, (III) plastic region, and (IV) failure region. The toe region represents the alignment of the collagen fibers. At 2% tension, all fibers are already out of their crimped state. In the linear region, the collagen fibers respond to the load in a linear fashion. The two subsequent regions (plastic and failure) represent the beginning and the total failure of the collagen fibers (accumulation of microdamages) (adapted from Robi et al. [39]).

A typical stress-strain curve has four distinctive regions. The first region is called *toe region*, where the crimped/zigzag pattern disappears when the strain of the tendon is below 2% and reappears when tension is released. Following the toe region, there is a linear region, in which the strain is lower than 4% (the physiological upper limit of strain in tendons) and the collagen fiber bundles are no longer crimped. If the strain remains lower than 4%, the fibers and fibrils have been shown to recoil back to their normal resting state, but strain greater than 4% can produce a microscopic damage (**Figure 2**). The slope of this region is Young's modulus, representing tendon stiffness. As the strain on the fibrils continues, the gap between the molecules increases and macroscopic failure can occur, with strain beyond 8–10%, eventually leading to tendon rupture [24, 38].

3. General aspects of tendon mechanical properties in response to exercise

Similar to the skeletal muscle, the tendon has the capacity to adapt according to their mechanical environment. In general, human movement occurs through force-generated muscle contraction, which is transmitted to the bone by aponeurosis and tendon [40]. On the other hand, movement is also essential for tendons [33, 41]. Since 1977, Beckham and coworkers have already noticed incomplete formation of cheek tendons during embryogenesis when muscle contraction was inhibited [42]. This result showed us that the relationship between the tendon and mechanical stimulus could be essential for tendon survival. Currently, we know the tendon answers metabolically to mechanical stimulus, and this exists in humans, for instance, tendon stiffness increase is related to mechanical loading imposed over it following a period of enhanced mechanical loading [33]. Moreover, short-term bout, as well as long-term loading-bearing, produces elevated collagen synthesis response [40]. A possible mechanism that explains the structural changes noticed following mechanical loading is the tendon responsiveness to mechanotransduction, promoting the interaction between fibroblasts and ECM. It is believed that this interaction between fibroblasts and ECM permits the cells to sense and respond to mechanical stimuli, promoting intracellular signaling that will improve protein synthesis and, consequently, tendon structures through collagen and growth factor autocrine/paracrine release (**Figure 3**).

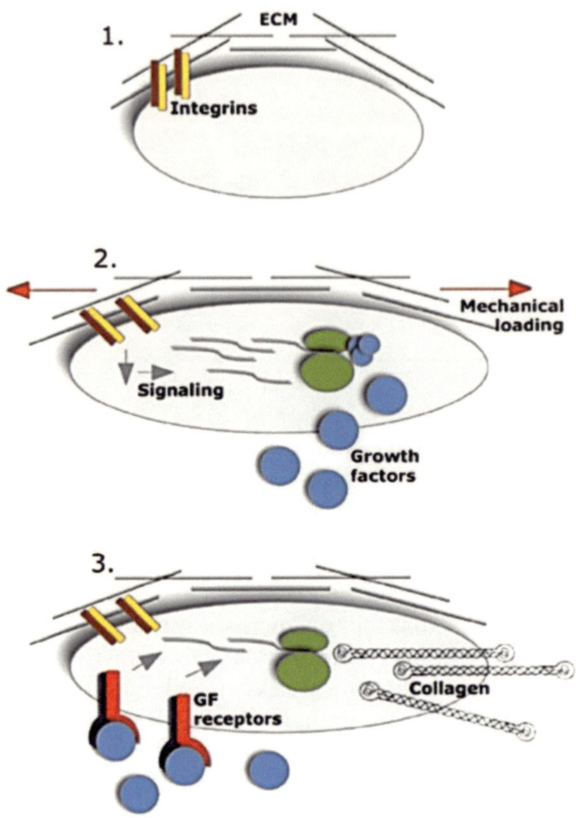

Figure 3.
Possible mechanism for loading induced collagen synthesis: (1) fibroblast connected to extracellular matrix via integrins. (2) Transcription and synthesis of growth factors induced by mechanical loading via changed intracellular signaling. (3) Autocrine-paracrine action of growth factors leading to increased collagen transcription and synthesis. Adapted from Heinemeier and Kjaer [43].

4. Biomechanical adaptations of the tendon in response to exercise training

Evidence-based interventions that elucidate biomechanical adaptations in the tendon are valuable to monitor the effectiveness of training programmers, as well as for the development, evaluation, and implementation of effective intervention programs aimed at injuries. The technique improvement for biomechanical evaluation in the tendon has directly influenced sports medicine, ergonomics, manufacturing sports equipment, and many other aspects of human life. Exercise training potentiates an increase of tensile strength, energy absorption, and stiffness, which may help to enhance the tendon quality. Several reports have already described that these alterations in tendon mechanical properties, due to changes in the loading levels, can improve the muscles' operating range and, consequently, the athletic performance [43].

4.1 Stiffness

Tendon stiffness is a crucial component for human locomotion and athletic performances because it keeps strain energy storage and returns properly and facilitates the muscle force potential due to force-length-velocity relationship [44]. Stiffness properties permit tendon to receive more or less stress, and this is directly related to tendon size. Thus, tendon stiffness is characterized by change in tendon length (ΔL) (deformation) in relation to the force applied to the tendon (ΔFt), but this parameter is dependent on the cross-sectional area (CSA) and length of the tendon, for instance, greater CSA and shorter length (deformation) will lead to greater stiffness [3] (**Figure 4**).

It has been demonstrated that tendon stiffness can increase after exercise training to maintain physiological ranges of strain. Tendon has been shown to undergo mechanical changes in response to diverse training regimens, including resistance and endurance training. The stiffness may vary according to limb dominance and specific activity. For example, patellar tendon stiffness was greater in the lead leg of badminton and fencing athletes [40]. The changes of the tendon material and morphological properties are among the prime candidate mechanisms, which could account for an increase of tendon stiffness in response to exercise. Running exercises can improve the tendon stiffness. Investigation that used ultrasonography reports greater tendon stiffness in long-distance runners than sedentary subjects [45]. The tendon stiffness may be a potential parameter for improving athletic performance in running, such as peak ground reaction force and ground contact time. The recent findings suggest that stiffer tendon may help achieve better running performance, with greater running economy, in endurance runners [46]. The sprinters had higher normalized stiffness (relation between tendon force and tendon strain) of the triceps sural tendon than the endurance runners and subjects not active in sports, which suggest that higher muscle strength possibly increases the margin of tendon tolerated mechanical loading due to tendon greater stiffness [47].

The resistance training (RT) potential for modulate tendon stiffness in individuals with connective tissue disorders, healthy individuals, athletes, and the elderly are largely described [48–50]. Eccentric exercise, isometric and plyometric training, and vascular occlusion are commonly used as forms of loading exercise for increasing tendon stiffness and represent important strategies for training and rehabilitation [51–53]. However, adding RT to endurance training did not affect stiffness patellar tendon compared to endurance training only [54]. The effects of RT on tendon stiffness can be potentiated by training variable manipulation, such as exercise intensity and duration program [33]. Several studies demonstrated that higher muscle contraction intensity (i.e., 70% of RM) showed higher stiffness values than low-intensity exercise (30% of RM), which indicates that exercise intensity exerts important regulation on tendon adaptation following mechanical loading exercise, regardless the muscle contraction type.

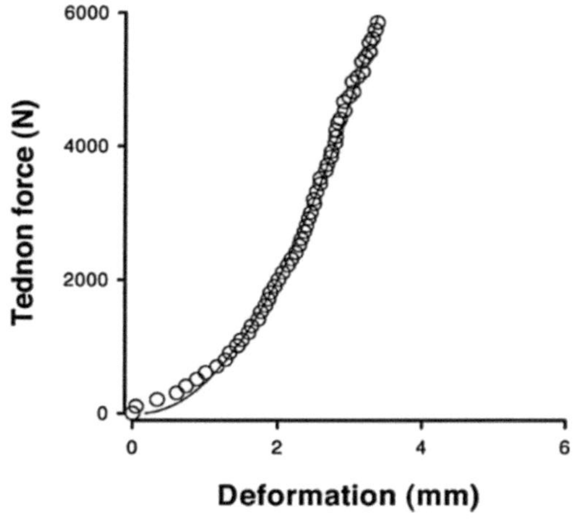

Figure 4.
Patella tendon force-derformation curve. Relationship between the force applied to the tendon (Ft) and the tendon deformation (elongation (ΔL). Adapted from Heinemeier and Kjaer [43].

In relation to the duration of the exercise program, several investigations suggest significant adaptations of tendon stiffness with only 8 weeks of intervention [33]; however, reduced elongation/strain over the whole force range can occur only after years of overload, indicating that there is a force-/strain-specific time course to these adaptations [55]. Although the use of RT may predict important biomechanical adaptations induced by training, the link between changes of tendon stiffness with different rest intervals, exercise order, and training volume remains to be determined.

On the other hand, aging can harm tendon biomechanical properties directly, as a result of biological change degeneration in the tendon and indirectly due to inactivity. In this context, studies suggest that RT is a therapeutic approach to minimize the deleterious effects of aging on biomechanical parameters in the tendon. In a recent review, the study suggests that the interventions should implement high mechanical loads with repetitive loading for up to 3–4 months to counteract age-related changes in muscle-tendon unit biomechanical properties [56]. Exercise training has demonstrated to promote stiffness [44] and increase the elastic modulus in elderly individuals [57]. Increased tendon stiffness is associated with a more rapid development of joint torque (~25%), which may be beneficial in the elderly. In rodent models, RT in old rats was effective for an increase in the stress-strain relationship, which improved the tendon capacity to support stress [36].

4.2 Tensile strength

The tendons need to be strong enough to sustain high magnitudes of loading, while their mechanical properties must remain functionally adequate for optimal muscle shortening or elastic energy storage. The tendons' tensile strength represents the pull which a tendon can resist without rupturing. This biomechanical parameter is a key measure to evaluate tendon injury risk [1]. It is worth highlighting that tendon injuries are extremely common, with the top three being tears of the rotator cuff, Achilles, and wrist flexor tendon. Exercise training has been able to produce excellent outcomes in about 75% of cases, with increased tensile strength being a key component. Several factors can affect the mechanical forces on tendons during locomotion and exercise [58]. In general, different tendons in the body are subject to different levels of mechanical loads, and both muscle contraction level and tendon relative size influence mechanical forces on a tendon.

In the past, the studies about exercise effects on tensile strength were based on in vitro animal investigations [59]. Recent advanced practices allow for noninvasive, in vivo assessment of fascicle movement and cross-sectional area of human tendons. The different athletic activities induce distinct levels of force, even on the same tendon. The tensile strength of a tendon is dependent on collagen and many proteoglycans, which proportionate viscoelastic properties, possessing both solid and fluid-like characteristics and exhibiting changes to the stress-strain relationship regarding the rate at which they are loaded [58]. This dynamic entity remodels permanently in response to mechanical stimulation. Most studies with heavy resistance exercise and endurance activity seem to indicate that systematic exercise strengthens the tendon complex [33].

There is a paradox regarding the exercise effects on tensile strength since acute exercise can induce a decrease in this biomechanical parameter, which may be detrimental to tendon functions. However, chronic exercise stands out as a remarkable non-pharmacological strategy for increasing tensile strength. It has been demonstrated that Australian football athletes' tendons presented a disorganization response to the mechanical loading of a game and that this disorganization returned to normal only after 4 days [60]. In the same research line, the football players showed an improvement in tendon structure over a 5-month preseason training period with increased fibrillar alignment [61]. Thus, exercise seems to improve tendon mechanical quality, including tensile strength. Tendons are able to withstand considerable forces during exercise, adapting to changes in mechanical load over time. Athletes of different modalities demonstrated improved viscoelastic properties, such as greater maximum tendon strain and suitable strain fluctuations when compared to nonathletes [62].

5. Tendon morphological properties in response to exercise training

Tendons' morphological properties are crucial to their ability to function effectively in situations of greater demand, such as the exercise training. The measures are determined from both the geometrical form and material properties. In general, the technological advances for tendon elongation evaluation, by means of an ultrasound-based methodology, as well as the determination of the tendon CSA from magnetic resonance images, enabled more robust elucidations. Although morphology analysis appear to be relatively simple measurements, researchers still encounter several difficulties that must be taken into account and that still limit current techniques [63].

5.1 Cross-sectional area

The tendons need to adapt based on the ratio of their peak operating stress to their ultimate stress. In this aspect, the modulation in CSA is an adaptive mechanism required for keeping suitable safety factors in response to larger stress levels. Several studies in vivo reported that tendon stiffening after exercise training is accompanied by increases in CSA, which indicated that tendon hypertrophy can also occur with mechanical loading. In a recent systematic review and meta-analysis about human tendon adaptation in response to mechanical loading, it was demonstrated that exercise training induces positive effect for CSA, regardless of type of applied loading regimens. Accumulating evidences from animal and human studies suggest increases in tendon CSA following exercise interventions. In these studies an increase of 20–30% of CSA was noticed in athletes that performed weight-bearing exercise (running and jumping) compared to athletes that did not performed weight-bearing exercise (kayakers) [64, 65]. Interestingly, a tendon CSA increase of 30% in athletes that performed sports where one lower extremity is normally submitted to more loading (leading leg) than the contralateral side was also observed.

In addition, Konsgaard et al. [66] have showed that 12 weeks of heavy RT in healthy young men were efficient to promote increase in both quadriceps tendon CSA in middle and distal regions, as well as in stiffness when compared to other leg that accomplished light RT. Therefore, it is known that loading magnitude could interfere in tendon size. Interestingly, athletes with tendon degeneration cannot present pain symptoms after tendon damage has occurred. It has been demonstrated that around 52% of distance runners will sustain an Achilles tendon injury during their career. In the context, it is extremely essential to determine the difference between tendon remodeling and tendon degeneration. Tendon CSA changes may further indicate positive exercise adaptations or initial degeneration and tendon tissue repair [67].

On the other hand, Wiesinger et al. investigated the tendon structural integrity in athletes engaged in sports with contrasting requirements [68]. Curiously, researchers showed that tendon CSA area normalized to body mass was smaller in water polo players than in other athletes (patellar and Achilles tendon, −28 to −24%) or controls (patellar tendon only, −9%). In contrast, the normalized cross-sectional area was larger in runners (patellar tendon only, +26%) and ski jumpers (patellar and Achilles tendon, +21% and + 13%, respectively) than in controls, which indicate that tendon morphological properties can be modulated by functional requirements.

At last, some studies have demonstrated that there are differences in tendon plasticity between men and woman. It was noticed that men's tendons hypertrophy with training, whereas trained women's tendons are the same size as those of untrained women. This work suggests that gender-specific humoral factors may be involved in the training-induced adaptive morphological response of the human tendon. In fact, tendon adaptation inhibition following exercise when levels of estrogen are high was observed [3, 10].

5.2 Tendon elongation

Tendon elongation correlates significantly with clinical outcome, and lengthening is an important cause of morbidity and may produce permanent functional impairment [69]. Increases in tendon stiffness and tensile strength reported following exercise training could conceivably be attributed to tendon elongation. Nevertheless, current literature does not offer conclusive evidence to support this premise. Several research lines about the positive effects of exercise on tendon elongation are controversial [33, 45]. Few studies considered all relevant methodological aspects (e.g., accounting for gravitational forces, axes misalignment of joint and dynamometer, averaging multiple trials to reliably assess tendon elongation, measuring the tendon moment arm directly). Possibly these aspects affect the validity of the applied method. In addition, whenever variations between measured and calculated tendon elongations are observed, it should be standard practice to confirm that there are no shortcomings in the original elongation calculations or the standard stressing procedures.

On the other hand, in tendon injury mechanical theory, tendon tissue overload is blamed for the pathologic process. Sports injuries, such as Achilles tendon rupture, are serious injuries for which the best treatment is still controversial. The main objective of intervention strategies must be to restore normal length, thus obtaining an optimal function [69]. Once tendon lengthening has become permanent, its clinical management is often difficult. The emphasis on exercise programs should be placed on muscle strengthening.

6. Tendon molecular signaling in response to exercise training

In general, the maintenance or changes in tendon CSA, as well as in tendon elongation, are regulated by interaction between synthesis and degradation molecular pathways [70–73]. In the tendon, the molecular adaptations stimulated by different types of exercise occur similarly to the skeletal muscle. Therefore, it is important to notice that in cases where training stimulates muscle hypertrophy and strength increase, the adaptations of muscles and tendons, which are collagen-rich tissues, are essential to maintain muscle-tendon unit integrity [74].

Tendons are distinct structures from muscles; however, tendon tissue has a direct continuation with the muscle ECM. This characteristic develops an essential mechanism that permits the communication of the mechanical properties of both muscles and tendons, allowing suitable force transmission between them [75]. Based on this communication, the externally applied mechanical load can stimulate ECM components through fibroblasts; however, ECM composition seems to be adapted specifically to changes in load. Therefore, it is possible to understand that mechanical stress can modulate the synthesis of ECM proteins, stimulating paracrine growth factor release, indirectly or directly triggering intracellular signaling pathway that will permit some ECM gene activation [76].

In order to investigate molecular signaling in response to exercise training, different approaches have been used, for instance, microanalysis, incorporation of stable isotope labeled proline into tendon tissue (C-13-proline), and mRNA gene expression of molecules present in the ECM. These approaches have been employed with the goal to analyze the modulations of several molecular mediators responsible for EMC remodeling, as well as molecules that present role key in tendon structure maintenance, such as collagens, proteoglycans, growth factors, as well as collagenases that could response to both endogenous and exogenous stimuli.

Currently, it is known that exercise induces collagen synthesis in the tendon, but the cellular mechanisms are still unclear. In the same way, growth factors as transforming growth factor-β-1 (TGF-β-1), connective tissue growth factor (CTGF), insulin-like growth factor (IGF), and mediator upstream involved in the collagen synthesis also might be involved in the ECM remodeling [77]. Enzymes involved in collagen processing, such as lysyl oxidase (LOX), in favor to cross-linking of collagen [78], as well as matrix metalloproteinases (MMPs) and their inhibitors (TIMPs) might contribute with tendon remodeling aggravated by exercise [79].

6.1 Collagens fibers and growth factors

The microanalysis has been employed to measure collagen pro-peptides which are responsible in the mature collagen synthesis.

Based on the microanalysis, it was possible to notice, in humans, elevated levels of collagen synthesis in peritendinous tissue in response to mechanical stimuli, in both acute and prolonged exercises [41]. These data corroborate with a previous study that used qPCR to investigate synthesized collagen by mRNA expression, but, in rats. In this study, the rats were submitted to 4 days of concentric, eccentric, or isometric training in the medial gastrocnemius muscle through sciatic nerve stimulation, simulating short-term strength training. Interestingly, in humans, high levels of type I and III collagen mRNA in the Achilles tendon in response in short-term resistance training were found, but no difference was seen between training types [77]. In this same study, researchers also investigated the regulation of TGF-β and CTGF key mediators for collagen fiber mRNA transcription.

TGF-β family is composed of more than 30 members identified in humans. This family orchestrates several cellular processes as proliferation, differentiation, protein metabolism, and growth and remodeling of the ECM in the tendon [80]. In the tendon, in particular, three TGF-β isoforms are known (TGF-β1, TGF-β2, and TGF-β3), but the most studied and that receive more attention is TGF-β1 isoform. In theory, latent TGF-β1 molecules are stored in ECM, in association with other ECM components such as fibrillin-1 and fibronectin. In cases when there is ECM remodeling necessity, for instance, injury tendon or overload following training,

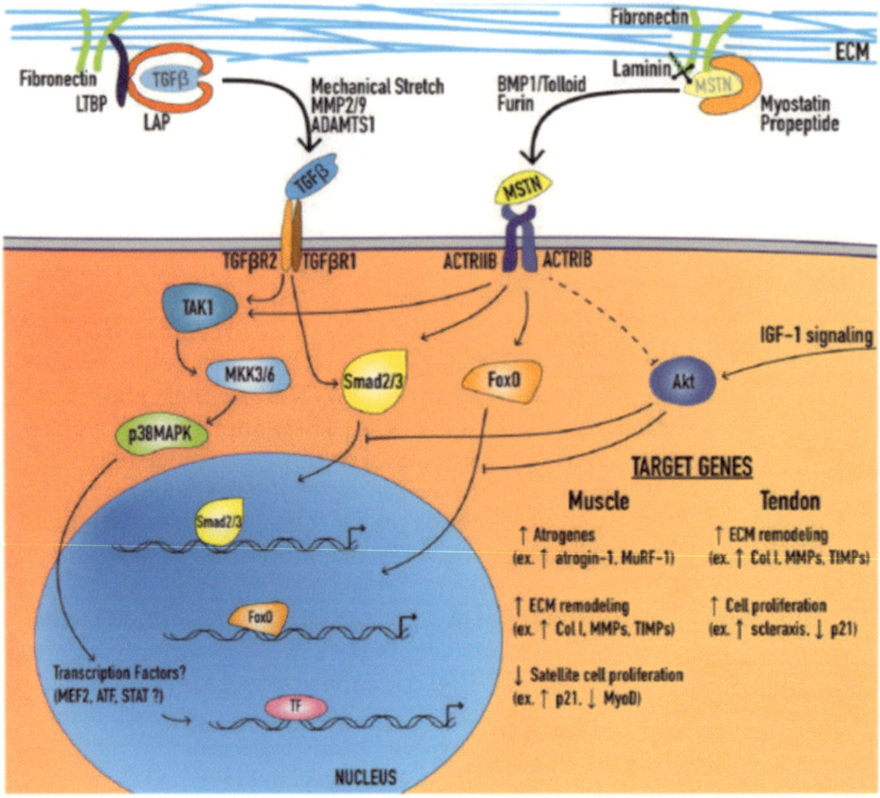

Figure 5.
Overview of TGF-β signaling pathways. Adapted from Gumucio et al. [75].

active TGF-β can be released of ECM through mechanical force or by matrix proteolytic enzymes as ADAMTS1, MMP-2, and MMP-9 [75]. In the case of mechanical force-mediated TGF-β activation, αvβ6 integrin, transmembrane proteins that connect intracellular cytoskeleton proteins together with ECM, suffers a conformational change that signals to liberate latent TGF-β; now matrix biologically active to binding in surface receptors are found in ECM cell. The binding between ligand (TGF-β) and their receptor permits activation of downstream intracellular signaling pathways, responsible for gene transcription, essential to ECM remodeling (for instance, collagen, MMPs, and TIMPs) (**Figure 5**) [75].

Interestingly, Heinemeyer et al. [77] confirmed in their study that TGF-β could be involved in collagen I and collagen III regulations in different types of training (concentric, eccentric, or isometric). Following 24 hours post training, a TGF-β increased gene expression in all types of training (concentric, eccentric, or isometric) with no difference among training types was noticed. These results are in accordance with previous studies that showed eccentric training is also accompanied by fibroblast proliferation, main cells responsible for synthesizing collagen in response to exercise [81, 82].

About ECM, connective tissue growth factor (CTGF), downstream mediator of TGF-β, in fibroblastic cells, also seems to be responsible for tendon ECM remodeling by exercise. It was noticed in human patellar tendon submitted to 1 hour of unilateral kicking exercise (workload of 67%) with frequency of 35 kicks per min and 2100 concentric contractions that CTGF gene expression total volume was increased, together with COL1A1 mRNA levels, 24 hours postexercise. On the other hand, tendon collagen protein synthesis between trained and untrained groups was not modified [83]. Despite that literature is still unclear about the mechanism that links mechanical loading, TGF-β-1, and CTGF, some results have reported that habitual loading is firstly related to stimulating proximal and distal portions of the tendon [40, 66]. However, future studies are necessary to obtain a better understating about the effect of exercise in this gene expression. Another point is inconsistency among loading protocol, so it is possible that some protocols have not reached to threshold enough stress-strain to stimulate the CTGF expression.

6.2 Proteoglycans

Other molecules, such as proteoglycans, are essential for fibrillogenesis regulation and tendon structure maintenance [84]. The proteoglycan regulation from exercise is still not clear on the literature, whereas most studies have observed the exercise effects over collagen and some growth factors responsible for gene expression of those molecules. However, it seems that resistance exercise appears not to induce changes in proteoglycan gene regulation. In the previous study, there were no observed changes in mRNA expression of the proteoglycans: decorin, biglycan, fibromodulin, and versican from resting levels at 4 or 24 hours after resistance training that corresponded to workload of 70% of the subject's concentric maximum repetition [84]. Although, this study hasn't found changes between proteoglycans, it is possible to infer that the regulation of these molecules could be related to mode, duration, and intensity of the exercise.

6.3 Matrix metalloproteinases

The ECM surrounding tendon provides structural support, protection, and maintenance of the functional integrity. The modulation of ECM function is controlled by matrix metalloproteinases (MMPs) and tissue inhibitors of metalloproteinases (TIMPs). The MMPs constitute a large family of endopeptidase zinc dependent that modulates ECM functions, degrading their constituents, such as proteoglycans, collagen, fibronectin, laminin, and other proteins during normal and pathological tissue remodeling. It has been reported that gelatinases (MMP-2 and MMP-9) play an important role in the ECM turnover induced by tissue injury and exercise training [84].

In order to compare exercise types (concentric, eccentric, or isometric) over gene expression of MMPs, a previous study noticed that MMP-2 mRNA expression increases moderately in the tendon in concentric and isometric exercises. On the other hand, MMP inhibitor, TIMP-1, and TIMP-2 increased gene expressions in response to all training types [77]. These data suggest a self-regulatory mechanism in attempt to protect the ECM against a high degradation of ECM compounds. In humans, it was found that MMP-2 mRNA decreased significantly 4 hours posteriorly to resistance training but returned to resting levels 24 hours after exercise. The mRNA expression of TIMP-1 did not change 24 hours post-acute exercise.

Interestingly in rodent models, previous data has already shown that acute or chronic exercises upregulate MMP-2 activity in the tendon, which is considered substantial mechanisms to tendon adaptation [85]. In contrast, anabolic-androgenic steroid treatment strongly inhibited this activity. Thus, anabolic-androgenic steroid treatment (AAS) can impair tissue remodeling in animal's tendons undergoing physical exercise by downregulating MMP activity, thus increasing the potential for tendon injury [86].

In the same research line, it has been demonstrated that the effects of exercise training on tendon repair are not the same for different tendon types and tendon regions (distal, proximal, and intermediary). For example, Marqueti et al. [87], showed that the intermediate region of the trained animals with AAS supplementation differed from the proximal and distal regions. Moreover, trained animals with AAS supplementation decreased MMP-2 activity form in three regions of the calcaneal tendon (distal, proximal, and intermediary) but not on the deep flexor tendons. The results suggest that the differences in the response to exercise and AAS treatment are a result of distinct metabolism and recruitment of these tendon regions in the exercise program. In another study, Pereira et al. [85] investigated MMP-2 activity in different regions of the calcaneal tendon after RT in ovariectomized rats. The authors demonstrated that ovarian hormones modulate MMP-2 activity differently in proximal region when compared to distal region; however, acute and chronic RT promote sufficient local stimuli to increase total and active MMP-2. Furthermore, proximal region of the calcaneus tendon seems to be more sensitive than the distal region to both acute and chronic RT due to greater MMP-2 activity increase, even in the ovariectomy condition.

7. Conclusions

In summary, tendons are highly responsive to morphological, biochemical, and biomechanical modifications in response to exercise training. Those changes emphasize the importance of extracellular matrix investigation and its remarkable characteristics in this tissue type. With respect to mechanical loading, is well known that exercise exerts beneficial effects in distinct regions of tendons. However, tendon remodeling is not the same in different tendon regions concerning the same mechanical loading application. Also, muscle contraction intensity is a key element in tendon adaptive responses. Finally, accumulating evidence from animal and human studies suggests several beneficial effects of exercise on the tendon remodeling, which might contribute to clinical conditions and performance, as well as understanding the potential of mechanical loading in different types of exercise conditions.

Author details

Rita de Cassia Marqueti*, Ivo Vieira de Sousa Neto, Fabricio Reichert Barin and Gracielle Vieira Ramos
University of Brasilia, Distrito Federal, Brazil

*Address all correspondence to: marqueti@gmail.com

References

[1] Wang JH, Guo Q, Li B. Tendon biomechanics and mechanobiology—A minireview of basic concepts and recent advancements. Journal of Hand Therapy. 2012;**25**(2):133-140; quiz 41

[2] Benjamin M, Kaiser E, Milz S. Structure-function relationships in tendons: A review. Journal of Anatomy. 2008;**212**(3):211-228

[3] Roberts TJ, Azizi E. The series-elastic shock absorber: Tendons attenuate muscle power during eccentric actions. Journal of Applied Physiologyl (1985). 2010;**190**(2):396-404

[4] Konow N, Azizi E, Roberts TJ. Muscle power attenuation by tendon during energy dissipation. Proceedings of the Biological Sciences. 2012;**279**(1731):1108-1113

[5] Wilson A, Lichtwark G. The anatomical arrangement of muscle and tendon enhances limb versatility and locomotor performance. Philosophical Transactions of the Royal Society of London. Series B, Biological Sciences. 2011;**366**(1570):1540-1553

[6] Ackermann PW, Renström P. Tendinopathy in sport. Sports Health. 2012;**4**(3):193-201

[7] Riley G. The pathogenesis of tendinopathy. A molecular perspective. Rheumatology (Oxford). 2004;**43**(2):131-142

[8] Kannus P. Structure of the tendon connective tissue. Scandinavian journal of medicine & science in sports. 2000;**10**(6):312-320

[9] Kjaer M. Role of extracellular matrix in adaptation of tendon and skeletal muscle to mechanical loading. Physiological Reviews. 2004;**84**(2):649-698

[10] Shoulders MD, Raines RT. Collagen structure and stability. Annual Review of Biochemistry. 2009;**78**:929-958

[11] Depalle B, Qin Z, Shefelbine SJ, Buehler MJ. Influence of cross-link structure, density and mechanical properties in the mesoscale deformation mechanisms of collagen fibrils. Journal of the Mechanical Behavior of Biomedical Materials. 2015;**52**:1-13

[12] Couppe C, Hansen P, Kongsgaard M, Kovanen V, Suetta C, Aagaard P, et al. Mechanical properties and collagen cross-linking of the patellar tendon in old and young men. Journal of Applied Physiology (1985). 2009;**107**(3):880-886

[13] Snedeker JG, Gautieri A. The role of collagen crosslinks in ageing and diabetes—The good, the bad, and the ugly. Muscles, Ligaments and Tendons. 2014;**4**(3):303-308

[14] Kostrominova TY, Calve S, Arruda EM, Larkin LM. Ultrastructure of myotendinous junctions in tendon-skeletal muscle constructs engineered in vitro. Histology and Histopathology. 2009 May;**24**(5):541-550

[15] Charvet B, Ruggiero F, Le Guellec D. The development of the myotendinous junction. A review. Muscles, Ligaments and Tendons Journal. 2012;**2**(2):53-63

[16] Schweitzer R, Zelzer E, Volk T. Connecting muscles to tendons: Tendons and musculoskeletal development in flies and vertebrates. Development. 2010;**137**(17):2807-2817

[17] Reeves ND, Cooper G. Is human Achilles tendon deformation greater in regions where cross-sectional area is smaller? The Journal of Experimental Biology. 2017;**220**(Pt 9):1634-1642

[18] Miles CA, Sims TJ, Camacho NP, Bailey AJ. The role of the alpha2 chain in the stabilization of the collagen type I heterotrimer: A study of the type I homotrimer in oim mouse tissues. Journal of Molecular Biology. 2002;**321**(5):797-805

[19] Sharma P, Maffulli N. Biology of tendon injury: Healing, modeling and remodeling. Journal of Musculoskeletal & Neuronal Interactions. 2006;**6**(2):181-190

[20] O'Brien M. The anatomy of the Achilles tendon. Foot and Ankle Clinics. 2005;**10**(2):225-238

[21] Rees SG, Flannery CR, Little CB, Hughes CE, Caterson B, Dent CM. Catabolism of aggrecan, decorin and biglycan in tendon. The Biochemical Journal. 2000;**350**(Pt 1): 181-188

[22] Peffers MJ, Thorpe CT, Collins JA, Eong R, Wei TK, Screen HR, et al. Proteomic analysis reveals age-related changes in tendon matrix composition, with age- and injury-specific matrix fragmentation. The Journal of Biological Chemistry. 2014;**289**(37):25867-25878

[23] Abate M, Silbernagel KG, Siljeholm C, Di Iorio A, De Amicis D, Salini V, et al. Pathogenesis of tendinopathies: Inflammation or degeneration? Arthritis Research & Therapy. 2009;**11**(3):235

[24] Freedman BR, Sarver JJ, Buckley MR, Voleti PB, Soslowsky LJ. Biomechanical and structural response of healing Achilles tendon to fatigue loading following acute injury. Journal of Biomechanics. 2014;**47**(9):2028-2034

[25] Lavagnino M, Wall ME, Little D, Banes AJ, Guilak F, Arnoczky SP. Tendon mechanobiology: Current knowledge and future research opportunities. Journal of Orthopaedic Research. 2015;**33**(6):813-822

[26] Maeda E, Ye S, Wang W, Bader DL, Knight MM, Lee DA. Gap junction permeability between tenocytes within tendon fascicles is suppressed by tensile loading. Biomechanics and Modeling in Mechanobiology. 2012;**11**(3-4):439-447

[27] McNeilly CM, Banes AJ, Benjamin M, Ralphs JR. Tendon cells in vivo form a three dimensional network of cell processes linked by gap junctions. Journal of Anatomy. 1996;**189**(Pt 3): 593-600

[28] Franchi M, Ottani V, Stagni R, Ruggeri A. Tendon and ligament fibrillar crimps give rise to left-handed helices of collagen fibrils in both planar and helical crimps. Journal of Anatomy. 2010;**216**(3):301-309

[29] Mountain KM, Bjarnason TA, Dunn JF, Matyas JR. The functional microstructure of tendon collagen revealed by high-field MRI. Magnetic Resonance in Medicine. 2011;**66**(2):520-527

[30] Shim V, Fernandez J, Besier T, Hunter P. Investigation of the role of crimps in collagen fibers in tendon with a microstructually based finite element model. Conference Proceedings: Annual International Conference of the IEEE Engineering in Medicine and Biology Society. 2012;**2012**:4871-4874

[31] Screen H, Birk D, Kadler K, Ramirez F, Young M. Tendon functional extracellular matrix. Journal of Orthopaedic Research. 2015;**33**(6):793-799

[32] Kubo K, Kanehisa H, Fukunaga T. Effects of viscoelastic properties of tendon structures on stretch-shortening cycle exercise in vivo. Journal of Sports Sciences. 2005;**23**(8):851-860

[33] Bohm S, Mersmann F, Arampatzis A. Human tendon adaptation in response to mechanical loading: A systematic review and meta-analysis of

exercise intervention studies on healthy adults. Sports Medicine - Open. Dec 2015;**1**(1):7:1-18

[34] Zhang J, Wang JH. The effects of mechanical loading on tendons—An in vivo and in vitro model study. PLoS One. 2013;**8**(8):e71740

[35] Fukutani A, Kurihara T, Isaka T. Factors of force potentiation induced by stretch-shortening cycle in plantarflexors. PLoS One. 2015;**10**(6):1-12

[36] de Cassia Marqueti R, Almeida JA, Nakagaki WR, Guzzoni V, Boghi F, Renner A, et al. Resistance training minimizes the biomechanical effects of aging in three different rat tendons. Journal of Biomechanics. 2017;**53**:29-35

[37] Shepherd JH, Screen HRC. Fatigue loading of tendon. International Journal of Experimental Pathology. 2013;**94**(4):260-270

[38] Wren TAL, Yerby SA, et al. Mechanical properties of the human achilles tendon. Clinical Biomechanics. 2001;**16**(3):245-251

[39] Robi K, Jakob N, Matevz K, Matjaz V. The physiology of sports injuries and repair processes. In Current issues in sports and exercise medicine. IntechOpen. 2013;**2**:48-49

[40] Couppe C, Kongsgaard M, Aagaard P, Hansen P, Bojsen-Moller J, Kjaer M, et al. Habitual loading results in tendon hypertrophy and increased stiffness of the human patellar tendon. Journal of Applied Physiology. 2008;**105**(3):805-810

[41] Beckham C, Dimond R, Greenlee TK Jr. The role of movement in the development of a digital flexor tendon. The American Journal of Anatomy. 1977;**150**(3):443-459

[42] Reeves ND. Adaptation of the tendon to mechanical usage. Journal of Musculoskeletal & Neuronal Interactions. 2006;**6**(2):174-180

[43] Heinemeier KM, Kjaer M. In vivo investigation of tendon responses to mechanical loading. Journal of Musculoskeletal & Neuronal Interactions. 2011;**11**(2):115-123

[44] Maganaris CN, Narici MV, Reeves ND. In vivo human tendon mechanical properties: Effect of resistance training in old age. Journal of Musculoskeletal & Neuronal Interactions. 2004;**4**(2):204-208

[45] Buchanan CI, Marsh RL. Effects of exercise on the biomechanical, biochemical and structural properties of tendons. Comparative Biochemistry and Physiology. Part A, Molecular & Integrative Physiology. 2002;**133**(4):1101-1107

[46] Takahashi C, Suga T, Ueno H, Miyake Y, Otsuka M, Terada M, et al. Potential relationship between passive plantar flexor stiffness and sprint performance in sprinters. Physical Therapy in Sport. 2018;**32**:54-58

[47] Arampatzis A, Karamanidis K, Morey-Klapsing G, De Monte G, Stafilidis S. Mechanical properties of the triceps surae tendon and aponeurosis in relation to intensity of sport activity. Journal of Biomechanics. 2007;**40**(9):1946-1952

[48] Han SW, Lee DY, Choi DS, Han B, Kim JS, Lee HD. Asynchronous alterations of muscle force and tendon stiffness following 8 weeks of resistance exercise with whole-body vibration in older women. Journal of Aging and Physical Activity. 2017;**25**(2):287-294

[49] Moller MB, Kjaer M, Svensson RB, Andersen JL, Magnusson SP, Nielsen RH. Functional adaptation of tendon and skeletal muscle to resistance training in three patients with genetically verified classic Ehlers Danlos

Syndrome. Muscles, Ligaments and Tendons Journal. 2014;**4**(3):315-323

[50] Waugh CM, Korff T, Fath F, Blazevich AJ. Effects of resistance training on tendon mechanical properties and rapid force production in prepubertal children. Journal of Applied Physiology (1985). 2014;**117**(3):257-266

[51] Roig M, Macintyre DL, Eng JJ, Narici MV, Maganaris CN, Reid WD. Preservation of eccentric strength in older adults: Evidence, mechanisms and implications for training and rehabilitation. Experimental Gerontology. 2010;**45**(6):400-409

[52] Hirayama K, Iwanuma S, Ikeda N, Yoshikawa A, Ema R, Kawakami Y. Plyometric training favors optimizing muscle-tendon behavior during depth jumping. Frontiers in Physiology. 2017;**8**:16

[53] Kubo K, Komuro T, Ishiguro N, Tsunoda N, Sato Y, Ishii N, et al. Effects of low-load resistance training with vascular occlusion on the mechanical properties of muscle and tendon. Journal of Applied Biomechanics. 2006;**22**(2):112-119

[54] Vikmoen O, Raastad T, Seynnes O, Bergstrom K, Ellefsen S, Ronnestad BR. Effects of heavy strength training on running performance and determinants of running performance in female endurance athletes. PLoS One. 2016;**11**(3):e0150799

[55] Massey GJ, Balshaw TG, Maden-Wilkinson TM, Folland JP. Tendinous tissue properties after short- and long-term functional overload: Differences between controls, 12 weeks and 4 years of resistance training. Acta Physiologica (Oxford, England). 2018;**222**(4):e13019

[56] McCrum C, Leow P, Epro G, Konig M, Meijer K, Karamanidis K. Alterations in leg extensor muscle-tendon unit biomechanical properties with ageing and mechanical loading. Frontiers in Physiology. 2018;**9**:150

[57] Grosset JF, Breen L, Stewart CE, Burgess KE, Onambele GL. Influence of exercise intensity on training-induced tendon mechanical properties changes in older individuals. Age (Dordrecht, Netherlands). 2014;**36**(3):9657

[58] Maganaris CN. Tensile properties of in vivo human tendinous tissue. Journal of Biomechanics. 2002;**35**(8):1019-1027

[59] Viidik A. Elasticity and tensile strength of the anterior cruciate ligament in rabbits as influenced by training. Acta Physiologica Scandinavica. 1968;**74**(3):372-380

[60] Rosengarten SD, Cook JL, Bryant AL, Cordy JT, Daffy J, Docking SI. Australian football players' Achilles tendons respond to game loads within 2 days: An ultrasound tissue characterisation (UTC) study. British Journal of Sports Medicine. 2015;**49**(3):183-187

[61] Docking SI, Rosengarten SD, Cook J. Achilles tendon structure improves on UTC imaging over a 5-month pre-season in elite Australian football players. Scandinavian Journal of Medicine & Science in Sports. 2016;**26**(5):557-563

[62] Mersmann F, Bohm S, Schroll A, Marzilger R, Arampatzis A. Athletic training affects the uniformity of muscle and tendon adaptation during adolescence. Journal of Applied Physiologyl (1985). 2016;**121**(4):893-899

[63] de Carvalho KL, Silva PE, Castro J, Babault N, Durigan JLQ, de Cassia Marqueti R. Height, weight, and age predict quadriceps tendon length and thickness in skeletally immature patients: Letter to the editor. The American Journal of Sports Medicine. 2017;**45**(9):NP26

[64] Kongsgaard M, Aagaard P, Kjaer M, Magnusson SP. Structural Achilles

tendon properties in athletes subjected to different exercise modes and in Achilles tendon rupture patients. Journal of Applied Physiology. 2005;**99**(5):1965-1971

[65] Rosager S, Aagaard P, Dyhre-Poulsen P, Neergaard K, Kjaer M, Magnusson SP. Load-displacement properties of the human triceps surae aponeurosis and tendon in runners and non-runners. Scandinavian Journal of Medicine & Science in Sports. 2002;**12**(2):90-98

[66] Kongsgaard M, Reitelseder S, Pedersen TG, Holm L, Aagaard P, Kjaer M, et al. Region specific patellar tendon hypertrophy in humans following resistance training. Acta Physiologica. 2007;**191**(2):111-121

[67] Sponbeck JK, Perkins CL, Berg MJ, Rigby JH. Achilles tendon cross sectional area changes over a division I NCAA cross country season. Journal of Exercise Science & Fitness. 2017;**10**(8):1226-1234

[68] Wiesinger HP, Rieder F, Kosters A, Muller E, Seynnes OR. Are sport-specific profiles of tendon stiffness and cross-sectional area determined by structural or functional integrity? PLoS One. 2016;**11**(6):e0158441

[69] Maquirriain J. Achilles tendon rupture: Avoiding tendon lengthening during surgical repair and rehabilitation. The Yale Journal of Biology and Medicine. 2011;**84**(3):289-300

[70] Cohen S, Nathan JA, Goldberg AL. Muscle wasting in disease: Molecular mechanisms and promising therapies. Nature Reviews. Drug Discovery. 2015;**14**(1):58-74

[71] Egerman MA, Glass DJ. Signaling pathways controlling skeletal muscle mass. Critical Reviews in Biochemistry and Molecular Biology. 2014;**49**(1):59-68

[72] Gundersen K. Muscle memory and a new cellular model for muscle atrophy and hypertrophy. The Journal of Experimental Biology. 2016;**219**(Pt 2):235-242

[73] Schiaffino S, Dyar KA, Ciciliot S, Blaauw B, Sandri M. Mechanisms regulating skeletal muscle growth and atrophy. The FEBS Journal. 2013;**280**(17):4294-4314

[74] Kjaer M, Magnusson P, Krogsgaard M, Boysen Moller J, Olesen J, Heinemeier K, et al. Extracellular matrix adaptation of tendon and skeletal muscle to exercise. Journal of Anatomy. 2006;**208**(4):445-450

[75] Gumucio JP, Sugg KB, Mendias CL. TGF-beta superfamily signaling in muscle and tendon adaptation to resistance exercise. Exercise and Sport Sciences Reviews. 2015;**43**(2):93-99

[76] Chiquet M, Renedo AS, Huber F, Fluck M. How do fibroblasts translate mechanical signals into changes in extracellular matrix production? Matrix Biology. 2003;**22**(1):73-80

[77] Heinemeier KM, Olesen JL, Haddad F, Langberg H, Kjaer M, Baldwin KM, et al. Expression of collagen and related growth factors in rat tendon and skeletal muscle in response to specific contraction types. The Journal of Physiology. 2007;**582**(Pt 3):1303-1316

[78] Kagan HM, Li W. Lysyl oxidase: Properties, specificity, and biological roles inside and outside of the cell. Journal of Cellular Biochemistry. 2003;**88**(4):660-672

[79] Visse R, Nagase H. Matrix metalloproteinases and tissue inhibitors of metalloproteinases: Structure, function, and biochemistry. Circulation Research. 2003;**92**(8):827-839

[80] Massague J. TGFbeta signalling in context. Nature Reviews Molecular Cell Biology. 2012;**13**(10):616-630

[81] Kjaer M, Langberg H, Heinemeier K, Bayer ML, Hansen M, Holm L, et al. From mechanical loading to collagen synthesis, structural changes and function in human tendon. Scandinavian Journal of Medicine & Science in Sports. 2009;**19**(4):500-510

[82] Mendias CL, Gumucio JP, Bakhurin KI, Lynch EB, Brooks SV. Physiological loading of tendons induces scleraxis expression in epitenon fibroblasts. Journal of Orthopaedic Research. 2012;**30**(4):606-612

[83] Dideriksen K, Sindby AK, Krogsgaard M, Schjerling P, Holm L, Langberg H. Effect of acute exercise on patella tendon protein synthesis and gene expression. SpringerPlus. 2013;**2**(1):109

[84] Sullivan BE, Carroll CC, Jemiolo B, Trappe SW, Magnusson SP, Dossing S, et al. Effect of acute resistance exercise and sex on human patellar tendon structural and regulatory mRNA expression. Journal of Applied Physiology. 2009;**106**(2):468-475

[85] Pereira GB, Prestes J, Leite RD, Magosso RF, Peixoto FS, Marqueti Rde C, et al. Effects of ovariectomy and resistance training on MMP-2 activity in rat calcaneal tendon. Connective Tissue Research. 2010;**51**(6):459-466

[86] Marqueti RC, Parizotto NA, Chriguer RS, Perez SE, Selistre-de-Araujo HS. Androgenic-anabolic steroids associated with mechanical loading inhibit matrix metallopeptidase activity and affect the remodeling of the achilles tendon in rats. The American Journal of Sports Medicine. 2006;**34**(8):1274-1280

[87] Marqueti RC, Prestes J, Paschoal M, Ramos OH, Perez SE, Carvalho HF, Selistre-de-Araujo HS. Matrix metallopeptidase 2 activity in tendon regions: Effects of mechanical loading exercise associated to anabolic-androgenic steroids. European Journal of Applied Physiology. 2008;**104**:1087-1093

Chapter 5

Patellar Tendinopathy: "Jumper's Knee"

Mayur Nayak and Rahul Yadav

Abstract

"Patellar tendinopathy" is also known as "Jumper's knee" and is a common cause of impaired function in athletes who participate in sports that require jumping and running activities. The exact etiology of disease is still unknown and several theories have been postulated for its pathogenesis. It usually presents as anterior knee pain that is related to the sports activity and might lead to decreased sports participation. USG and MRI are the main modality of investigation that aids in the diagnosis. Non-operative therapy forms the main stay of treatment in form of rest, brace, physical therapy and anti-inflammatory medications. Other adjuncts such as cryotherapy, corticosteroids injection, platelet -rich plasma injections and electrical therapy like TENS or ESWT have been used with some success. Operative intervention in form of open or arthroscopic procedures are reserved for chronic and refractory cases.

Keywords: patellar tendinopathy, degenerative, tendinitis, jumper's knee, enthesis, eccentric, cryotherapy

1. Introduction

Tendinopathy is a broad term encompassing painful condition in and around tendon due to overuse. It commonly affects the Extensor mechanism (patellar tendon) around the knee and is termed as "jumper's knee" [1, 2]. It is characterized by an initial reactive or inflammatory response followed by stage of degeneration. The prevalence of patellar tendinosis is seen greatest in young adults who are engaged in high demand sports that involves running, jumping and cutting movements. It has been estimated to range from 40%-50% in elite volleyball players and 35%-40% in high level basketball players [3]. It can also affect sedentary individuals and an estimated prevalence of 14.2% is seen in general population [4].

The disease pass through two stages, acute and chronic and presentation vary according to the stage of disease. It usually present as anterior knee pain, swelling and impaired function in acute stage, while in chronic stage it may presents as long standing anterior knee pain with profound muscle wasting without the sign of inflammation or impaired function. The tendo-osseous junction on the inferior pole of patella is the usual affected site. The exact etiology of disease is still unknown. However, various risk factor have been identified which may contribute in development of patellar tendinopathy such as impaired quadriceps flexibility and strength, high body mass index (BMI), leg length discrepancy, impaired hamstring flexibility and vertical jump performance, as all these factors increases the strain over the patellar tendon [5]. Several theories have been postulated on pathogenesis of patellar tendinopathy including vascular [6], mechanical [7], nervous [8] and impingement related. However the chronic repetitive tendon overload theory is the most accepted theory for patellar tendinosis.

The term "patellar tendinitis" which is often used for patellar tendon pain appears to be misnomer for patellar tendinopathy. Multiple histo-pathological studies have reported that the primary pathology in most of painful tendon is degenerative rather than inflammatory [9–11]. However Fredberg et al. [12] challenged the concept of patellar tendon pain due to degenerative cause and stated that it is rather the presence of inflammatory process that is responsible for pain. While some histopathological studies [13, 14] have shown that pro-inflammatory chemical agent such as cyclooxygenase, growth factors, and prostaglandins are present in acute stages and macrophage and lymphocyte in chronic tendinopathy, we still need further research and more evidences to prove the theory.

The management of patellar tendinopathy has always been challenge to the healthcare professionals. It usually requires a multimodal approach and based upon the current literature and clinical practice an effective conservative intervention in the form of rest, NSAIDS and physiotherapy is indicated in acute phase and surgical procedure are reserved for chronic and long standing cases.

2. Anatomy

Patellar tendon is the continuation of the common tendon of insertion of quadriceps, extending from the inferior pole of patella to the tibial tuberosity. In adult, the patellar tendon is around 25–40 mm wide in the coronal plane and 4–5 mm deep in the sagittal plane and 4–6 cm long. Macroscopically it appears white, glistening and stringy with collagen fiber in tendon are arranged in parallel fashion.

The blood supply of the tendon is through the anastomotic vascular ring lying in the thin layers of loose connective tissue that covers the fibrous expansion of the rectus femoris. The formation of the ring is through the anterior tibial recurrent artery and genicular arteries mainly the lateral superior, lateral inferior and medial inferior artery [15, 16].

The patellar tendon attachment to bone (patella or tibia) has fibrocartilaginous enthesis with four distinct tissue zones–dense fibrous connective tissue, uncalcified fibrocartilage, calcified cartilage and bone [17]. The patellar tendon lack of well-developed proper paratenon, while the posterior surface of tendon which is in direct contact of fat pad, which is highly vascular and innervated. According to Duri et al. [18] the intensity of pain in some patient with patellar tendinopathy is due to involvement of fat pad. Patellar tendon pathology usually involve the enthesis site, it most commonly involve the inferior pole of patella but can also involve tibial tubercle or proximal aspect of patella in quadriceps tendon. Macroscopically the diseased portion of tendon become disorganized and appears yellow-brown in color.

3. Epidemiology

Jumper's knee is commonly seen in people involved in contact sports such as basketball, volleyball, high jump, long jump, tennis and running [1]. The factor contributing factor in development of patellar tendinopathy can be classified as extrinsic factor or intrinsic factor [19]. The intrinsic factor can be sex, race, bone structure, bone density, muscle length, muscle strength, joint range of motion. While extrinsic factors are training volume (duration, frequency and intensity), specific sports activity, specific movements like quick acceleration, deceleration, cutting action and training surfaces. Ferreti [20] observed direct relation between the patellar tendinopathy with number of weekly training session and training over the hard surface.

The condition is more commonly seen in males than females [21]. This condition is not a self-limiting one and symptoms may prevail after treatment.

4. Pathogenesis

Tendons display a classical stress-strain curve, during the increased flexion the maximum load is located in deep posterior portion of patella, close to center of rotation knee and inferior pole of patella. The crimp pattern of the tendon disappears when the length of the tendon is stretched greater than 2%. With further stretch greater than 5%, the tendon fibers become more parallel and the tendon follows a linear response to stress [22, 23]. Beyond this, tendon failure starts to begin with disruption of collagen cross links.

The force experienced by patellar tendon on a level ground while walking is 0.5 kN, which increases to 8 kN during landing from a jump, 9 kN during fast running and further to 14.5 kN during competitive weight lifting. A basketball player, on an average, jumps 70 times per game where the vertical component of the ground reaction force reaches to about six to eight times the body weight. Thus, sport activities can impose high levels of stress on the tendon, enough to cause its failure. This increased strain result in alteration of cellular activity level by affecting the tenocytes and altering the protein and enzyme production with deforming of nucleus [7, 24]. The tendon fibroblast are loaded with increased prostaglandins E2, leukotrienes B4, VEGF and matrix metalloproteinase which contribute to tendinopathy. This increased strain result in alteration of cellular activity level by affecting the tenocytes and altering the protein and enzyme production with deforming of nucleus [7, 24]. The tendon fibroblast are loaded with increased prostaglandins E2, leukotrienes B4, VEGF and matrix metalloproteinase which contribute to tendinopathy.

In chronic patellar tendinopathy, there will be absent or minimal inflammation [10, 13, 21]. The diseased tendon shows hypercellularity with atypical fibroblast and endothelial cell proliferation with neovascularization [13, 21, 25, 26]. There will be loss of longitudinal collagen fibers and demarcation between collagen bundles with relative expansion of tendon. There will be abundant number of cell undergoing apoptosis with abundance of pre-apoptotic proteins and gene [27, 28]. Macroscopically tendinopathy will have disorganized appearance described as mucoid degeneration [20, 29].

5. History and physical examination

The clinical diagnosis of jumper's knee is based on the subjective sensation of the pain and restriction of activities. The symptoms are insidious in onset but usually relates to an increase in frequency or intensity of activity involving rapid repetitive ballistic movements of the knee joint. It starts with a dull aching pain in the anterior aspect of the knee after a strenuous activity and further progresses to a state where it interferes with the performance of the individual. Patients also complain of pain while ascending or descending stairs.

The key physical finding of the patellar tendinopathy is tenderness at the inferior pole of patella when the knee is at full extension with gradual decrease in the pain when the knee is flexed gradually. It is generally accompanied by few signs of soft tissue inflammation [1]. The condition often associated with abnormalities of patellar tracking, chondromalacia patellae, Osgood-schlatter disease or mechanical malalignment of the leg.

6. Imaging

6.1 Roentgenogram

Anteroposterior, lateral and tangential views of patella may show radiolucency over involved pole in early stage, while in prolonged stage, the involved pole may appear elongated. In chronic cases tendon calcification and periosteal reaction over anterior aspect of patella ("tooth sign") may be evident. While in long standing cases the stress fracture or disruption of extensor mechanism may occur.

6.2 Ultrasonography

Ultrasound and MRI is the main modality of investigation for jumper's knee. Although CT scan can image patellar tendon, it does not offer any significant advantage over the above mentioned investigations. However studies have shown CT scan to be of some prognostic value [30, 31].

Role of ultrasound:

1. Detect preclinical lesion in athletes

2. Detect patellar tendon pathology and assess its severity

Ultrasonographic appearance:

Patellar tendons in patients suffering from jumper's knee have decreased echogenicity, containing either a sonolucent region or diffuse hypoechogenicity [31–33]. The tendon envelope is irregular, and there may be erosion of patellar tip and intratendinous calcification may be present.

6.3 Role of MRI

1. Identification of exact location and the extension of the tendon involvement

2. Exclusion of other condition such as bursitis, chondromalacia

3. Quantification of the size of patellar tendon to be excised during surgery

Appearance in MRI:

The abnormal patellar tendon contains an oval to round area of high signal intensity in T1- and T2-sequence at the tendon attachment site or focal zones of high signal intensity in the deeper zones of the tendon [34, 35]. The T2-weighted sequences (particularly the T2*-weighted GRE sequences) have greater sensitivity than the T1-weighted protocols. However, the T1-weighted signal can image most cases of patellar tendinopathy.

Ferrati et al. [36] classified jumper's knee into six stages according to symptoms:

Stage	Symptoms
0	No pain
1	Pain only after intense sports activity, no undue functional impairment
2	Pain at the beginning and after sports activity, still able to perform at a satisfactory level
3	Pain during sports activity, increasing difficulty in performing at a satisfactory level
4	Pain during sports activity, unable to participate in sports at as satisfactory level
5	Pain during daily activity, unable to participate in sport at any level

7. Management

Unfortunately the management used for chronic tendon disorders has a very little scientific backdrop and varies considerably among the surgeons and across countries. Thus the treatments listed below are at best empirical.

7.1 Rest

This modality is useful in athletes presenting for the first time with patellar pain but becomes a concern in individuals involved in competitive sports.

7.2 Straps, braces and exercises

One approach to help in healing of the tendon tissues is unloading. This can be achieved either by the modification of the activities for e.g. decreasing the number of jumps or landing on the ground with proper orientation of the foot, or use of knee braces and straps such as chopart straps that share and alter the load on the tendon. Chopart strap [10] is a tape attached just proximal to tibial tendon. The most popular non-operative treatment involves eccentric exercise.

7.3 Cryotherapy and physical modalities

Cryotherapy helps in controlling initial tissue response to the injury. It is thought to act by decreasing the blood flow and metabolic rate, thereby decreasing the rate of inflammation. Electrical modalities that have been used in patellar tendinopathy include ESWT, ultrasound, heat, interferential therapy, magnetic fields, pulsed magnetic and electromagnetic fields, transcutaneous electrical nerve stimulation (TENS), and laser [10, 37, 38]. The true effects of the above mentioned modalities still remain unknown and further studies are required to support their use.

7.4 Remedial massage

It tends to treat the tendon tissue by having an effect on the muscle stretch and direct effect on the tendon cells. Muscle belly massage is thought to increase the compliance of the muscle and decrease the load on the muscle. Deep friction massage is thought to activate the mesenchymal cells to stimulate the healing response. "Fibrolysis", a form of deep frictional massage originally developed in Finland, has been successful in Achilles tendinopathy. However a controlled study failed to provide any evidence of healing in patellar tendinopathy and further evidence is required to warrant its use [39].

7.5 Rehabilitation

The key treatment nowadays to chronic tendinopathy is the stretching and strengthening programme of the whole muscle tendon unit. A staged program for tendinopathy corresponding to the stages of worsening severity of the condition [1] was outlined by Stanish et al. [40] outlined in **Table 1**.

Drop squat forms one of the key exercise for this condition in which patients are asked to sit to about 100–120° of knee flexion from a standing position and are advised to perform three sets consisting of 10 repetitions per session. It was observed that this regimen brought about complete relief in 30% of patients with reduction in the symptoms in further 64% of the patients [41]. Worsening symptoms were seen in the remaining 6% of patients. Cannell [42] observed that eccentric squats were better as compared to leg curl/extension exercises in treatment of the condition.

Stage	Program
Stage 1	Adequate warm up Ice after activity Local anti-inflammatory treatments, including non-steroidal anti-inflammatory drugs (NSAIDs) Physiotherapy(isometric quadriceps exercises and an elastic knee support)
Stage 2	Addition of period of rest and heat before activity
Stage 3	Addition of prolonged period of rest

Table 1.
Program for patellar tendinopathy.

7.6 NSAIDS

Although the benefits of NSAIDs are dubious, they are the most common drug used for symptomatic relief [43]. Although the use of "anti-inflammatory" medication seems paradoxical for a condition that is essentially degenerative, it is believed NSAIDs might act via mechanisms different from their conventional anti-inflammatory actions [44]. In vitro studies in human cartilage have revealed a variable interaction of NSAIDs with glycosaminoglycans (GAGs) where some have been shown to stimulate and some to inhibit, its synthesis [45]. This mechanism also sheds light on its effect on the synthesis of extra-cellular matrix. In a double blinded placebo controlled study, the use of NSAIDs in tendinopathy, piroxicam did not benefit patients with Achilles tendinopathy however topical ketoprofen reached the target tissue in patients with patellar tendinopathy, but the clinical efficacy was not assessed [46].

7.7 Corticosteroids

Corticosteroids are known for reducing the symptoms arising from the inflamed synovial structures. However the role of corticosteroids remains controversial in management of tendinopathy. According to Jozsa and Kannus [47] steroids are contraindicated in acute phase of tendinopathy and in the late chronic phase of tendinopathy when the tendon degeneration is advanced which may lead to tendon rupture. However it has proved beneficial when diluted with anesthetic for diagnostic reasons and to minimize adverse effects and in conditions where 1–6 week rest period combined with a programme of gradual strengthening is required before returning to activity.

7.8 Other medical treatments

Aprotinin, an 85 amino acid 65 kDa basic polypeptide extracted from bovine lungs has shown to offer better pain relief than steroids at least in short term. However aprotinin which is a strong inhibitor MMP (matrix metalloproteinase) is less effective in insertional tendinopathy as compared to main body [48]. Another non-surgical treatment option includes use of sclerosing agent with chemical irritant (e.g. polidocanol) [8, 49, 50]. These targets the neovascularization and accompanying nerves. The use of platelet-rich plasma injection has been tried in tendinopathy and favorable outcome have been found [51–53]. However still there is no level 1 or level II studies about role of PRP in patellar tendinopathy. The glyceryl trinitrate (GTN) patch [54, 55], which delivers nitric oxide (NO) to pathological tendon which play role in tendon healing. But we still need level1or level II evidence to support it.

7.9 Surgical treatment

Patellar tendon surgery is indicated in patients who have failed conservative management more than 6 weeks [56–58]. A variety of surgical procedures have been described such as resection of the tibial attachment of the patellar tendon with realignment, drilling of the inferior pole of the patella, macroscopic necrotic area excision [59], repair of macroscopic defects, longitudinal tenoplasty/tenotomy of the tendon [60] percutaneous longitudinal tenotomy, percutaneous needling [61] and arthroscopic assisted decompression [62, 63] of the tendon, possibly with excision of the inferior pole of the patella however the effectiveness of any single procedure has not been elucidated.

8. Conclusions

Patellar tendinopathy is essentially a degenerative condition and the management should be based on the clinical assessment. Imaging appearances, although aids in the diagnosis but should not determine the treatment. Conservative treatment forms the mainstay of management, while surgery is indicated only after a dedicated period of appropriate conservative measures have been instituted, usually around 6–9 months. These include physical modalities such as local application of ice and graduated strengthening physiotherapy protocol such as functional exercises and eccentric strengthening; the latter are done only after the patient is pain free. Although there is a lack of level I evidence, eccentric training appears to be the most promising modality. Peritendinous corticosteroid or aprotinin infiltration may also be useful as an adjunct for the treatment of this condition. Although scientific consensus is lacking percutaneous needling appears to be the least invasive procedure, followed by percutaneous longitudinal tenotomy. Arthroscopic debridement has been proposed, but, although early results are encouraging, its efficacy is still under scrutiny.

Conflict of interest

No conflict of interest.

Acronyms and abbreviations

BMI	body mass index
NSAID	nonsteroidal anti-inflammatory drug
VEGF	vascular endothelial growth factor
MRI	magnetic resonance imaging
CT	computed tomography
GRE	gradient echo sequence
ESWT	extracorporeal shockwave therapy
TENS	transcutaneous electrical nerve stimulation
MMP	matrix metalloproteinase
PRP	platelet rich plasma
GTN	glyceryl tri-nitrate
NO	nitric oxide

Author details

Mayur Nayak* and Rahul Yadav
All India Institute of Medical Sciences, New Delhi, India

*Address all correspondence to: mayur.nayak86@gmail.com

References

[1] Blazina ME, Kerlan RK, Jobe FW, Carter VS, Carlson GJ. Jumper's knee. The Orthopedic Clinics of North America. 1973;**4**(3):665-678

[2] Depalma MJ, Perkins RH. Patellar tendinosis: Acute patellar tendon rupture and jumper's knee. The Physician and Sportsmedicine. 2004;**32**(5):41-45

[3] Lian OB, Engebretsen L, Bahr R. Prevalence of jumper's knee among elite athletes from different sports: A cross-sectional study. The American Journal of Sports Medicine. 2005;**33**(4):561-567

[4] Zwerver J, Bredeweg SW, van den Akker-Scheek I. Prevalence of Jumper's knee among nonelite athletes from different sports: A cross-sectional survey. The American Journal of Sports Medicine. 2011;**39**(9):1984-1988

[5] van der Worp H, van Ark M, Roerink S, Pepping G-J, Van den Akker-Scheek I, Zwerver J. Risk factors for patellar tendinopathy: A systematic review of the literature. British Journal of Sports Medicine. 2011;**45**(5):446-452

[6] MacAuley D. Do textbooks agree on their advice on ice? Clinical Journal of Sport Medicine. 2001;**11**(2):67

[7] Magra M, Maffulli N. Genetic aspects of tendinopathy. Journal of Science and Medicine in Sport. 2008;**11**(3):243-247

[8] Alfredson H, Öhberg L. Neovascularisation in chronic painful patellar tendinosis—Promising results after sclerosing neovessels outside the tendon challenge the need for surgery. Knee Surgery, Sports Traumatology, Arthroscopy. 2005;**13**(2):74-80

[9] Alfredson H, Lorentzon R. Chronic tendon pain: No signs of chemical inflammation but high concentrations of the neurotransmitter glutamate. Implications for treatment? Current Drug Targets. 2002;**3**(1):43-54

[10] Khan KM, Cook JL, Bonar F, Harcourt P, Astrom M. Histopathology of common tendinopathies. Update and implications for clinical management. Sports Medicine (Auckland, N.Z.). 1999;**27**(6):393-408

[11] Maffulli N. Overuse tendon conditions: Time to change a confusing terminology. Arthroscopy: The Journal of Arthroscopic & Related Surgery. 1998;**14**(8):840-843

[12] Fredberg U. Tendinopathy—Tendinitis or tendinosis? The question is still open. Scandinavian Journal of Medicine Science in Sports. 2004;**14**(4):270-272

[13] Fu SC, Wang W, Pau HM, Wong YP, Chan KM, Rolf CG. Increased expression of transforming growth factor-beta1 in patellar tendinosis. Clinical Orthopaedics. 2002;**400**:174-183

[14] Rolf CG, Fu BS, Pau A, Wang W, Chan B. Increased cell proliferation and associated expression of PDGFRbeta causing hypercellularity in patellar tendinosis. Rheumatology (Oxford, England). 2001;**40**(3):256-261

[15] Scapinelli R. Blood supply of the human patella. Its relation to ischaemic necrosis after fracture. Journal of Bone and Joint Surgery. British Volume (London). 1967;**49**(3):563-570

[16] Scapinelli R. Studies on the vasculature of the human knee joint. Acta Anatomica (Basel). 1968;**70**(3):305-331

[17] Benjamin M, Kumai T, Milz S, Boszczyk BM, Boszczyk AA, Ralphs JR. The skeletal attachment

of tendons—Tendon "entheses". Comparative Biochemistry and Physiology. Part A, Molecular & Integrative Physiology. 2002;**133**(4):931-945

[18] Duri ZA, Aichroth PM, Wilkins R, Jones J. Patellar tendonitis and anterior knee pain. The American Journal of Knee Surgery. 1999;**12**(2):99-108

[19] Reinking MF. Current concepts in the treatment of patellar tendinopathy. International Journal of Sports Physical Therapy. 2016;**11**(6):854-866

[20] Ferretti A. Epidemiology of jumper's knee. Sports Medicine (Auckland, N.Z.). 1986;**3**(4):289-295

[21] Cook JL, Khan KM, Harcourt PR, Grant M, Young DA, Bonar SF. A cross sectional study of 100 athletes with jumper's knee managed conservatively and surgically. The Victorian Institute of Sport Tendon Study Group. British Journal of Sports Medicine. 1997;**31**(4):332-336

[22] Structure and function of mammalian tendon [Internet]. 2018. Available from: https://www.researchgate.net/publication/9243172_Structure_and_function_of_mammalian_tendon

[23] Hess GP, Cappiello WL, Poole RM, Hunter SC. Prevention and treatment of overuse tendon injuries. Sports Medicine. 1989;**8**(6):371-384

[24] Magra M, Maffulli N. Matrix metalloproteases: A role in overuse tendinopathies. British Journal of Sports Medicine. 2005;**39**(11):789-791

[25] Rees JD, Maffulli N, Cook J. Management of tendinopathy. The American Journal of Sports Medicine. 2009;**37**(9):1855-1867

[26] Popp JE, Yu JS, Kaeding CC. Recalcitrant patellar tendinitis. Magnetic resonance imaging, histologic evaluation, and surgical treatment. The American Journal of Sports Medicine. 1997;**25**(2):218-222

[27] Benson RT, McDonnell SM, Knowles HJ, Rees JL, Carr AJ, Hulley PA. Tendinopathy and tears of the rotator cuff are associated with hypoxia and apoptosis. Journal of Bone and Joint Surgery. British Volume (London). 2010;**92**(3):448-453

[28] Lian Ø, Scott A, Engebretsen L, Bahr R, Duronio V, Khan K. Excessive apoptosis in patellar tendinopathy in athletes. The American Journal of Sports Medicine. 2007;**35**(4):605-611

[29] Fritschy D, Wallensten R. Surgical treatment of patellar tendinitis. Knee Surgery, Sports Traumatology, Arthroscopy. 1993;**1**(2):131-133

[30] King JB, Perry DJ, Mourad K, Kumar SJ. Lesions of the patellar ligament. Journal of Bone and Joint Surgery. British Volume (London). 1990;**72**(1):46-48

[31] Mourad K, King J, Guggiana P. Computed tomography and ultrasound imaging of jumper's knee-patellar tendinitis. Clinical Radiology. 1988;**39**(2):162-165

[32] Maffulli N, Regine R, Carrillo F, Minelli S, Beaconsfield T. Ultrasonographic scan in knee pain in athletes. British Journal of Sports Medicine. 1992;**26**(2):93-96

[33] Teitz CC. Ultrasonography in the knee. Clinical aspects. Radiologic Clinics of North America. 1988;**26**(1):55-62

[34] el-Khoury GY, Wira RL, Berbaum KS, Pope TL, Monu JU. MR imaging of patellar tendinitis. Radiology. 1992;**184**(3):849-854

[35] Khan KM, Bonar F, Desmond PM, Cook JL, Young DA, Visentini PJ, et al.

Patellar tendinosis (jumper's knee): Findings at histopathologic examination, US, and MR imaging. Victorian Institute of Sport Tendon Study Group. Radiology. 1996;**200**(3):821-827

[36] Ferretti A, Conteduca F, Camerucci E, Morelli F. Patellar tendinosis: A follow-up study of surgical treatment. The Journal of Bone and Joint Surgery. American Volume. 2002;**84-A**(12):2179-2185

[37] Lee EW, Maffulli N, Li CK, Chan KM. Pulsed magnetic and electromagnetic fields in experimental achilles tendonitis in the rat: A prospective randomized study. Archives of Physical Medicine and Rehabilitation. 1997;**78**(4):399-404

[38] van der Worp H, van den Akker-Scheek I, van Schie H, Zwerver J. ESWT for tendinopathy: Technology and clinical implications. Knee Surgery, Sports Traumatology, Arthroscopy. 2013;**21**(6):1451-1458

[39] Pellecchia GL, Hamel H, Behnke P. Treatment of infrapatellar tendinitis: A combination of modalities and transverse friction massage versus iontophoresis. Journal of Sport Rehabilitation. 1994;**3**:135-145

[40] Stanish WD, Curwin S, Rubinovich M. Tendinitis: The analysis and treatment for running. Clinics in Sports Medicine. 1985;**4**(4):593-609

[41] el Hawary R, Stanish WD, Curwin SL. Rehabilitation of tendon injuries in sport. Sports Medicine (Auckland, N.Z.). 1997;**24**(5):347-358

[42] Cannell LJ. The effects of an eccentric-type exercise versus a concentric-type exercise in the management of chronic patellar tendonitis [Internet]. University of British Columbia; 1982. Available from: https://open.library.ubc.ca/cIRcle/collections/ubctheses/831/items/1.0077278

[43] Rolf C, Movin T, Engstrom B, Jacobs LD, Beauchard C, Le Liboux A. An open, randomized study of ketoprofen in patients in surgery for Achilles or patellar tendinopathy. The Journal of Rheumatology. 1997;**24**(8):1595-1598

[44] Weiler JM. Medical modifiers of sports injury. The use of nonsteroidal anti-inflammatory drugs (NSAIDs) in sports soft-tissue injury. Clinics in Sports Medicine. 1992;**11**(3):625-644

[45] Dingle JT, Leeming MR, Martindale JJ. Effect of tenidap on cartilage integrity in vitro. Annals of the Rheumatic Diseases. 1993;**52**(4):292-299

[46] Åström M, Westlin N. No effect of piroxicam on achilles tendinopathy. A randomized study of 70 patients. Acta Orthopaedica Scandinavica. 1993;**63**:631-634

[47] Paavola M, Kannus P, Järvinen TAH, Järvinen TLN, Józsa L, Järvinen M. Treatment of tendon disorders. Is there a role for corticosteroid injection? Foot and Ankle Clinics. 2002;**7**(3):501-513

[48] Khan KM, Maffulli N, Coleman BD, Cook JL, Taunton JE. Patellar tendinopathy: Some aspects of basic science and clinical management. British Journal of Sports Medicine. 1998;**32**(4):346-355

[49] Alfredson H, Öhberg L. Sclerosing injections to areas of neo-vascularisation reduce pain in chronic Achilles tendinopathy: A double-blind randomised controlled trial. Knee Surgery, Sports Traumatology, Arthroscopy. 2005;**13**(4):338-344

[50] Hoksrud A, Bahr R. Ultrasound-guided sclerosing treatment in patients with patellar tendinopathy (jumper's knee). 44-month follow-up. The

American Journal of Sports Medicine. 2011;**39**(11):2377-2380

[51] Kon E, Filardo G, Delcogliano M, Presti ML, Russo A, Bondi A, et al. Platelet-rich plasma: New clinical application: A pilot study for treatment of jumper's knee. Injury. 2009;**40**(6):598-603

[52] Bowman KF, Muller B, Middleton K, Fink C, Harner CD, Fu FH. Progression of patellar tendinitis following treatment with platelet-rich plasma: Case reports. Knee Surgery, Sports Traumatology, Arthroscopy (KSSTA). 2013;**21**(9):2035-2039

[53] Gosens T, Den Oudsten BL, Fievez E. Pain and activity levels before and after platelet-rich plasma injection treatment of patellar tendinopathy: A prospective cohort study and the influence of previous treatments. International Orthopaedics. 2012;**36**(9):1941-1946

[54] Steunebrink M, Zwerver J, Brandsema R, Groenenboom P, van den Akker-Scheek I, Weir A. Topical glyceryl trinitrate treatment of chronic patellar tendinopathy: A randomised, double-blind, placebo-controlled clinical trial. British Journal of Sports Medicine. 2013;**47**(1):34-39

[55] Paoloni JA, Appleyard RC, Nelson J, Murrell GAC. Topical nitric oxide application in the treatment of chronic extensor tendinosis at the elbow: A randomized, double-blinded, placebo-controlled clinical trial. The American Journal of Sports Medicine. 2003;**31**(6):915-920

[56] Karlsson J, Kälebo P, Goksör LA, Thomée R, Swärd L. Partial rupture of the patellar ligament. The American Journal of Sports Medicine. 1992;**20**(4):390-395

[57] Karlsson J, Lundin O, Lossing IW, Peterson L. Partial rupture of the patellar ligament. Results after operative treatment. American Journal of Sports Medicine. 1991;**19**(4):403-408

[58] Yu JS, Popp JE, Kaeding CC, Lucas J. Correlation of MR imaging and pathologic findings in athletes undergoing surgery for chronic patellar tendinitis. AJR. American Journal of Roentgenology. 1995;**165**(1):115-118

[59] Orava S, Osterback L, Hurme M. Surgical treatment of patellar tendon pain in athletes. British Journal of Sports Medicine. 1986;**20**(4):167-169

[60] Martens M, Wouters P, Burssens A, Mulier JC. Patellar tendinitis: Pathology and results of treatment. Acta Orthopaedica Scandinavica. 1982;**53**(3):445-450

[61] Leadbetter WB, Mooar PA, Lane GJ, Lee SJ. The surgical treatment of tendinitis. Clinical rationale and biologic basis. Clinics in Sports Medicine. 1992;**11**(4):679-712

[62] Lee DW, Kim JG, Kim TM, Kim DH. Refractory patellar tendinopathy treated by arthroscopic decortication of the inferior patellar pole in athletes: Mid-term outcomes. The Knee. 2018;**25**(3):499-506

[63] Lorbach O, Diamantopoulos A, Paessler HH. Arthroscopic resection of the lower patellar pole in patients with chronic patellar tendinosis. Arthroscopy: The Journal of Arthroscopic & Related Surgery. 2008;**24**(2):167-173

Chapter 6

Mechanisms of Action of Multipotent Mesenchymal Stromal Cells in Tendon Disease

Janina Burk

Abstract

Multipotent mesenchymal stromal cells (MSCs) are a promising therapeutic tool to treat tendon disease. Aiming to establish successful treatment approaches and to fully exploit the regenerative potential of the MSC, it is crucial to understand their mechanisms of action. However, these can be multifaceted and strongly context-sensitive and are still not well-understood in the context of tendon disease. This review aims to shed light on the different possible mechanisms, including engraftment, tenogenic differentiation, extracellular matrix synthesis and remodeling, immunomodulation, pro-angiogenetic effects, trophic support, and protection of resident tendon cells. Evidence from experimental and clinical (veterinary) case studies was compiled and interpreted in conjunction with the respective in vitro and animal models used.

Keywords: MSC, ASC, progenitor cell, tendon, mechanism of action, engraftment, differentiation, extracellular matrix, remodeling, immunomodulation, trophic support

1. Introduction

Tendinopathy is a common cause of recurring pain and long-term impairment in leisure and professional athletes, increased age being an additional risk factor. The prevalence of clinically manifest conditions in risk groups is high: in a cohort of football players, 21% suffered from Achilles tendon problems [1]. Moreover, even in clinically healthy volunteers, ultrasonographic evidence of Achilles tendon alterations was found in 16% [2]. This indicates that clinical manifestation is only the tip of the iceberg, the basis of which is a long-term interplay of inflammatory and degenerative changes.

Tendons have to withstand high mechanical loads and serve as an energy storage with elastic properties. The required biomechanical properties are provided by the extracellular matrix (ECM) [3], which is largely composed of hierarchically structured, cross-linked, and crimped collagen type I fibrils. The tenocytes, while representing only 5% of the tissue volume, maintain the ECM structure by constant remodeling. This normally enables biochemical and biomechanical adaptations to exercise [4]. Recurrent overuse impairs this physiological adaptation.

The onset of tendinopathy is currently understood as the result of a failed healing response to repeated tissue trauma. Microruptures, oxidative, mechanical, and heat stress activate resident cells and trigger a cascade of inflammation

IntechOpen

and degeneration, culminating in ECM deterioration. Key molecules involved include vascular endothelial growth factor (VEGF), interleukin (IL)-1, tumor necrosis factor (TNF)-α, prostaglandin (PG)E2, glutamate, and substance P [5, 6]. These mediators foster the ingrowth of blood vessels and nerves and the activation of nociceptive pathways. They are also implicated in the upregulation and activation of matrix metalloproteinases (MMP) and downregulation of their endogenous inhibitors (tissue inhibitors of matrix metalloproteinases; TIMP) [7]. This entails ECM degradation which successively alters and weakens the ECM structure [6]. When the accumulated damage and sensitization reach a threshold, clinical manifestation of tendinopathy comprises classical signs of inflammation including pain. Furthermore, provoked by new overload events, massive tissue trauma can occur. The resolution of inflammation is crucial to limit tissue damage, yet this mechanism often fails. Promoting fibrosis, a lack of pro-resolving signals, and persistence of macrophages entails the continuing activation of fibroblasts [8, 9]. Furthermore, macrophages could further contribute to ECM degradation via MMP secretion. Once at a diseased state, the intrinsic regenerative capacity of tendons is poor. Although endogenous mesenchymal stem-like cells with high tenogenic potential reside within tendons [10–12], these are susceptible to damage and suffer age-related changes [13, 14]. In pathological states, they could even contribute to fatty degeneration, fibrosis, and heterotopic ossifications [15, 16].

Treatment of tendinopathy still represents an unsolved challenge. Mainly, the use of strict rehabilitation exercise regimens is sufficiently evidence based [17, 18]. Anti-inflammatory drugs are frequently used, but they do not only counteract the active inflammation but also its resolution [19]. Biologicals such as platelet rich plasma have also received much attention, but clinical evidence is not convincing [17, 20, 21]. Research also focuses on the potential of endogenous tendon progenitor cells [22], which may be a promising strategy but will not be addressed in this review.

Multipotent mesenchymal stromal cells (MSCs) represent a therapeutic tool which might meet the clinical need of an adaptive treatment that simultaneously addresses different aspects of the disease. MSCs reside in virtually any tissue, in close proximity to the vasculature [23, 24]. MSCs derived from bone marrow and adipose tissue (BMSC and ASC, respectively) have been most extensively characterized [25, 26]. The fibroblast-like cells have been defined by a set of inclusion and exclusion antigens, their plastic-adherence, and trilineage differentiation potential in vitro [26]. While their differentiation potential into mesenchymal cell types, including tenocytes [27], has led to their extensive use in tissue engineering, it has become evident that their therapeutic potential by far exceeds cell replacement [24, 28]. While proof of MSC engraftment is often lacking, MSC-based cell therapy has shown beneficial effects in diverse scenarios in animal models, mostly mediated by immunomodulatory and trophic mechanisms [29–33]. Particularly, the immunomodulatory potential is extensively being researched and already exploited clinically, e.g., for treatment of graft-versus-host disease [34–36].

The use of MSC for tendon repair was first suggested in 1998 [37] and, interestingly, has been published as a case report on an equine patient as early as 2003 [38]. Since then, several experimental animal studies—the recent ones being reviewed here—and case series in equine patients [39–41] have raised hope that local implantation of MSC into acute tendon defects improves healing. However, translational progress into human orthopedics is underwhelming, and although equine patients are being treated and few first-in-man clinical trials have been performed or initiated [42–44], convincing evidence from randomized, controlled clinical studies has neither been obtained in equine nor in human patients so far [45]. This may in part be due to our still limited understanding of the MSC mechanisms of action in tendon healing, which delays the development of targeted treatment approaches.

The aim of this review was to collect the evidence for the different possible MSC mechanisms of action in the treatment of tendon disease. In vitro and in vivo studies published within the last 5 years were screened and their results were compiled, focusing on MSC-based cell therapy using BMSC or ASC.

2. Tendon regeneration and defect models

2.1 In vitro and ex vivo models

In vitro and ex vivo models relevant to MSC mechanisms of action in tendon regeneration comprise two major groups, with some overlap (**Figure 1**). The first includes the wide range of models for tenogenic differentiation [10, 46–94]. Among these, approaches in three-dimensional dynamic cultures appear most representative for MSC mechanisms in vivo [57, 58, 64, 70, 74, 77, 79, 83, 84, 86, 87]. Typically assessed parameters following tenogenic differentiation include the expression of tenogenic transcription factors (scleraxis and, in the more recent studies, mohawk), the transmembrane glycoprotein tenomodulin, as well as the expression and deposition of extracellular matrix components (e.g., collagen I, collagen III, decorin, and tenascin-C) and biomechanical parameters in case of tissue engineered constructs. Upregulation of matrix components such as collagen I or tenascin-C and improved construct strength do not only suggest tenogenic differentiation but also indicate ECM-modulating activities of the MSC. However, it should be acknowledged that no truly specific tendon marker has yet been identified, and that only expression patterns of combined marker sets, e.g., collagen I, scleraxis, and tenascin-C, discriminate healthy tendon from diseased tendon or other musculoskeletal tissues [95].

The second group includes models investigating the interaction of MSC with tenocytes and/or the tendon ECM, using co-cultures of MSC and tenocytes, their respective conditioned media, or tendon explants [48, 69, 74, 75, 88, 91, 92, 94, 96–105].

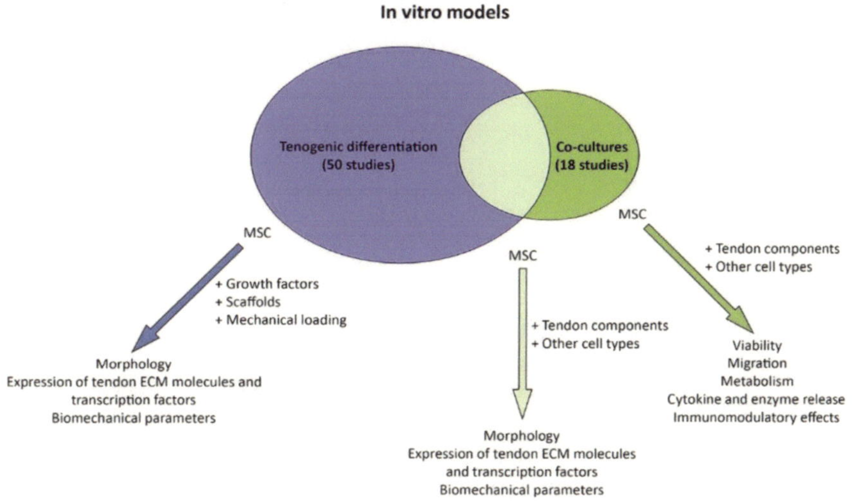

Figure 1.
In vitro models.

Outcome parameters assessed in these studies are more diverse and include cell viability, proliferation, and metabolic parameters, expression and/ or release of growth factors, cytokines, MMPs and TIMPs, expression of ECM receptors and cytoskeleton formation, ECM protein release or deposition, or modulatory effects on immune cells (e.g., macrophage M1/M2 switch). Consequently, these studies provide insight into MSC trophic effects, immunomodulatory, or matrix-modulatory mechanisms.

The figure gives an overview of the in vitro models included in this review, illustrating the overlap between tenogenic differentiation models and coculture models, and summarizes the most commonly assessed outcome parameters. Note that in this context, the term "coculture" is used to summarize the models investigating the interplay between tenocytes and MSC, thus it does not exclusively refer to cocultures of different cell types but also includes cell culture models using conditioned media or tendon explants.

2.2 In vivo models

In vivo studies on MSC-based tendon therapies need to be discriminated with respect to the animal model used (small vs. large, type of disease or defect model) and the treatment approach (strategy for MSC delivery, possible adjuvant treatments, timing of treatment, MSC source, and cell numbers applied).

Animal species used comprise small (rats [54, 106–118] and rabbits [119–122]) and large animals (dogs [123–126], sheep [127–129], and horses [130–141]). Interestingly, there appears to be a fair balance between small and large animal studies. This suggests preclinical progress, but it is also due to the interest in the equine species within the veterinary community. The tendon defects were created surgically in the majority of studies, with full thickness transections or segmental defects (mostly in the Achilles tendon) used in small animals or dogs and surgically created core lesions in the superficial digital flexor tendons in the equine model. Although there is reason to believe that enzymatical induction of tendon lesions better mimics the ECM degeneration and inflammation in tendon disease, only few among the recent studies used collagenase-based tendinopathy models [106, 108, 110, 129, 137, 139]. Still, neither surgical nor enzyme-based approaches fully reflect the complex tendon pathophysiology. In this light, providing particularly valuable information, some studies in the equine species were performed using horses suffering from naturally occurring tendinopathy [134, 138, 141] (**Figure 2**).

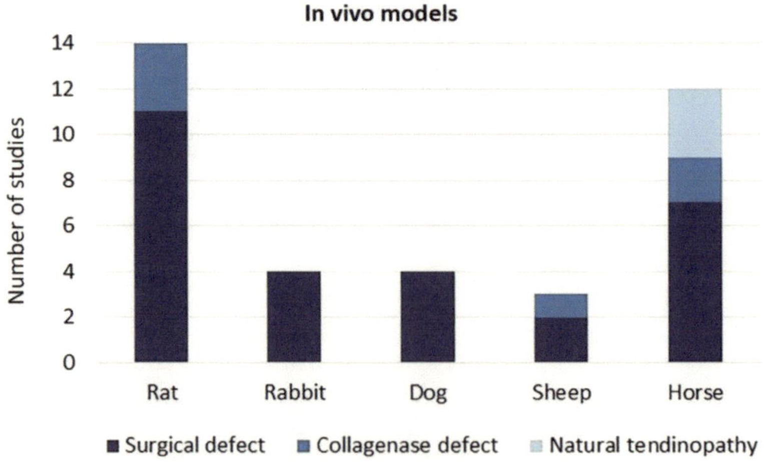

Figure 2.
In vivo models.

The diagram displays the numbers of studies performed in different animal species which were included in this review and indicates the types of tendon defect models used in the respective species.

Approaches for MSC implantation include local delivery of MSC suspensions, mostly via (ultrasound-guided) injection [106–112, 119, 120, 127–133, 136–141], coating of suture materials with MSC [113], MSC delivery in fibrin-based vehicles [54, 114, 124] or cell sheets [54, 123, 125, 126], and the use of diverse constructs of MSC and scaffold materials [115–118, 121, 122]. Interestingly, while the delineation between MSC-scaffold constructs for MSC delivery and for tendon replacement is sketchy, it is remarkable that construct-based approaches are almost exclusively used in small animals. This indicates that translational progress using these approaches is poor, possibly due to their incapability to meet the biomechanical demands in large animals or humans.

Further aspects of the treatment approach are likely to influence MSC mechanisms of action and complicate the coherent interpretation of findings from different studies. Adjuvant treatments, e.g., simultaneous growth factor delivery, or pre-treatment of the MSC, such as pre-differentiation or inflammatory licensing before cell delivery, may support certain mechanisms synergistically but may negatively interfere with other mechanisms. For example, bone morphogenetic protein (BMP)-12 promotes MSC tenogenic differentiation but reduces their immunomodulatory potential [93]. Next, the timing of the treatment is of great importance as different mechanisms of action of MSC are likely to be relevant during different stages of tendon healing. Furthermore, the dosage, i.e., the numbers of MSC applied, may not only play a role with respect to treatment efficacy but also with respect to supporting specific mechanisms of action [120]. For example, interactions between MSC and immune cells depend on the ratio of MSC to leukocytes present [142].

Last not least, the MSC source is likely to influence their mechanisms of action, which is an issue with equal relevance for in vitro findings. On the one hand, this applies to the choice of donor in terms of age and health status [143] and in terms of autologous, allogeneic or, in case of many small animal models, even xenogeneic use of MSC. On the other hand, the tissue origin of MSC as well as the donor species impact on the cell characteristics [57, 140, 144] and thus potentially on their mechanisms of action. Therefore, mainly studies focusing on the well-characterized BMSC and ASC were included and their tissue origin discriminated where appropriate. Furthermore, it was attempted to compile only studies which enabled the discrimination of MSC effects from those of possible additional treatments. In this line, in vivo studies using genetically engineered MSC for other purposes than cell tracking were not included in this review.

3. Engraftment and tenogenic differentiation

The assumption that MSC engraftment and their tenogenic differentiation after implantation into a tendon lesion lead to the replacement of damaged tenocytes dates back to the earlier days of MSC research and mirrors the general conception of MSC at that time [27, 38]. In the following years, the fact that MSC persistence at the site of tissue damage could not be achieved in models for a wide variety of diseases led to the assumption that differentiation and cell replacement might not even contribute to the regenerative effects observed after MSC transplantation [28]. This hypothesis was fostered by the compelling finding that paracrine factors released by the MSC can lead to similar beneficial effects as the MSC themselves, leading to the concept of cell-free MSC-based therapies [145]. Still, the situation might be slightly different in tendon pathologies, and at the moment, it cannot be excluded that tenogenic differentiation of engrafted cells could contribute to regeneration, perhaps as a basis for further trophic and ECM-modulatory mechanisms.

3.1 In vitro evidence

An extensive body of recent literature describes the tenogenic differentiation of MSC in response to a wide range of stimuli, although unfortunately, no generally accepted in vitro model or standard tenogenic differentiation assay exists. Current concepts of tenogenic differentiation are reviewed in detail elsewhere [146, 147]. The most commonly used stimuli to induce tenogenesis in MSC include growth factors, scaffolds with specific topography, and cyclic mechanical loading, with most studies combining two or more of these approaches, based on earlier studies in the field of tissue engineering [37, 148–150].

Growth factors used for induction of tenogenic differentiation mainly include transforming growth factor-β family members (TGF-β [47, 51, 53, 60, 66, 86, 88] and the growth differentiation factors GDF-5/BMP-14 [60, 67, 68, 70, 82, 151], GDF-6/BMP-13 [72], GDF-7/ BMP-12 [56, 60, 80, 93], and GDF-8 [71, 78]) but also fibroblast growth factors (FGF) [49, 89, 90], insulin-like growth factor-1 [53], vascular endothelial growth factor (VEGF) [60], or epidermal growth factor [49]. A promising stepwise differentiation approach has also been reported using TGF-β1 followed by connective tissue growth factor (CTGF) [54]. Growth factors are commonly delivered as culture medium supplements, but, e.g., FGF-2-transduced MSCs have been used as well [89]. Further tenogenic differentiation approaches based on genetic modifications include the forced expression of the tenogenic transcription factors scleraxis [10, 152] or mohawk [52, 116].

Currently used scaffolds comprise decellularized tendon matrices [57, 58, 64, 65, 83, 84, 88] and (synthetic) scaffolds with specifically designed topography and stiffness [59, 61–63, 68, 70, 72–75, 79, 81, 87], both being used based on evidence that physical cues such as scaffold anisotropy and stiffness direct MSC fate. Decellularized tendon matrices provide biochemical cues at the same time. A different approach to exploit the natural tendon biochemical composition is to use tendon ECM or tenocytic extracts as a culture supplement [46, 47, 91].

Mechanical loading of cell cultures, typically MSC-seeded scaffolds, is performed in bioreactors, most commonly by uniaxial cyclic stretching [46, 57, 58, 64, 66, 70, 74, 77, 79, 83, 84, 86, 87]. Different frequencies and strain rates have been used. While results are consistent in that cyclic stretching supports tenogenic differentiation, there is a discrepancy regarding the extent of stretching, with some studies highlighting moderate strain rates of 2 or 3% as beneficial for tenogenic induction [58, 77], while others support the use of higher strain rates (e.g. 10%) [55, 153]. Further approaches to tenogenic differentiation by physical stimulation include the use of extracorporeal shock waves [76], pulsed electromagnetic fields [85], and the activation of mechanosensitive membrane receptors [50].

In addition to using growth factors, scaffolds, and mechanical loading, tenogenic differentiation of MSC has also been reported in co-cultures with tenocytes [48, 69, 74, 75, 92] or in tenocyte-conditioned medium [48].

This overview illustrates that a wide range of stimuli can induce a tenogenic phenotype in MSCs (BMSCs as well as ASCs), although the quality of differentiation cannot be directly compared between studies and certainly varies. With respect to possible MSC tenogenic differentiation in vivo, the studies relying on physiological stimuli, such as mechanical loading, biomimetic scaffolds, or cross-talk with tenocytes, are most insightful. In contrast, the use of growth factors (typically at concentrations exceeding those found in vivo) or genetic modifications is suitable for mechanistic studies and may be helpful for tenogenic pre-differentiation prior to MSC implantation but does not reflect the in vivo situation. To understand if physiological stimuli could promote the same distinct tenogenic phenotype as artificial TGF-β concentrations, it would be helpful to gain further insight into the downstream signaling networks and their possible interfaces. So far, however, tenogenic signaling has mainly been investigated following growth factor stimulation [67, 82, 89, 90]. Only few studies have attempted to elucidate the signaling pathways activated in MSC in response to mechanical load or scaffold topographical cues, focusing on the role of rho/ROCK [154, 155].

Yet, although physiological stimuli have repeatedly been shown to induce tenogenic differentiation in MSC, it should not be anticipated that this mechanism is analogously activated when MSCs are implanted into a tendon lesion. Self-evidently, the tendon lesion does not provide a physiological but rather a pathophysiological environment, which may have an entirely different impact on the MSCs. Unfortunately, this issue is still underrepresented in the current literature. Recently, we investigated ASC tenogenic properties in response to physiological tenogenic and simultaneous inflammatory stimulation [84]. This study demonstrated that ASC tenogenic properties are compromised not only in the presence of the pro-inflammatory cytokines IL-1β and TNF-α but also in the presence of leukocytes. Similarly, IL-1β and IL-6 inhibited tenogenic differentiation in tendon-derived stem cells [156, 157]. Furthermore, again in tendon-derived stem cells, stiff matrices impeded tenogenic differentiation [158]. Together, these findings suggest that MSC tenogenic differentiation may be impaired in a pathophysiological in vivo environment, which can comprise inflammatory stimuli as well as stiff (fibrotic) ECM, depending on the stage of disease.

3.2 In vivo evidence

Although extensively investigated in vitro, there is no distinctive evidence of tenogenic differentiation following MSC implantation in vivo. One conceivable explanation is that MSC differentiation is in fact impaired in the pathophysiological lesion environment. Nevertheless, in contrast to studies in other disease models, MSCs have been repeatedly localized in treated tendon lesions, providing a basis for long-term regenerative effects, possibly including differentiation and cell replacement. Furthermore, there is some evidence of homing of MSCs to tendon lesions, although not unambiguous. The mechanism of homing may be of minor importance with respect to cell delivery at the macroscale, as the cells are almost exclusively delivered locally in MSC-based tendon therapies. Yet, the capability of homing is still indicative of MSCs that are capable of identifying regions of tissue damage at the microscale, where they would actively integrate.

None of the small animal studies included in this review specifically addressed MSC homing to tendon lesions. However, when bursal tissue was implanted in rotator cuff tendon lesions in a rat model, the green fluorescent protein-labeled mesenchymal stem cells from this tissue infiltrated the healing tendons [159], demonstrating the presence of homing signals. Accordingly, ASC infiltration into the tendons was also evident when cell sheets were used as delivery vehicle in a canine model [126]. However, when injected into the tendon sheath, BMSC homed to synovial structures but were not attracted to the tendon lesions in an ovine model of intrasynovial tendon healing [127]. In the equine large animal model, homing of MSC to tendon lesions has been addressed in more detail. Scintigraphic short-term in vivo tracking of technetium-labeled BMSC showed that the cells homed to the tendon lesion after administration by regional limb perfusion, although local administration by direct intralesional injection was more effective, and no homing was observed after intravenous administration. These findings were consistent between artificial tendon lesions [135] and natural tendinopathy [134]. Interestingly, intraarterial limb perfusion showed greater accumulation of BMSC in the lesion on day 10 after surgical lesion induction than on day 3 [135]. This finding illustrates that the stage of tendon disease is of importance to MSC homing mechanisms. However, scintigraphic tracking also revealed that even after local injection, only a relatively small proportion of the injected BMSC remains at the injury site (24% after 24 h) [134]. In accordance with this, we and others demonstrated that ASCs are distributed via the bloodstream within the first few days after their injection into equine tendon lesions, possibly as they are washed away before they can home and attach [136, 139]. We additionally observed that the ASCs were subsequently also found in nontreated tendon lesions, indicating their capability of homing [139].

Engraftment of MSC within treated tendon lesions was demonstrated in several studies, albeit results are not conclusive as to the numbers of surviving cells in relation to the cell numbers administered. In rat Achilles tendon defects, BMSC or ASC could be identified histologically at 2, 4, and 8 weeks after cell implantation (injection) [107, 109, 112], as well as 3 weeks after implantation of a BMSC-seeded collagen scaffold [116]. Complementing these small animal studies, MSCs have been traced in large animal studies, including longitudinal in vivo cell tracking. In sheep, green or red fluorescent protein-labeled BMSCs were detected histologically at 1, 2, 3, 4, and 6 weeks following their implantation [128, 129]. In the equine model, we and others could trace superparamagnetic iron oxide-labeled ASC by magnetic resonance imaging during follow-up periods of up to 24 weeks after implantation into artificial tendon lesions [132, 139] and umbilical cord tissue-derived MSCs during a follow-up period of 8 weeks in naturally occurring tendinopathy [138]. In the experimental tendon lesions, histological results confirmed the presence of the simultaneously fluorochrome-labeled ASC until week 24 [132, 139]. This provides evidence for a remarkable long-term persistence of part of the locally injected MSC, yet it has neither been proved nor disproved whether these cells commit to a tenogenic fate.

4. Extracellular matrix modulation

The restoration of the ECM architecture and functionality is a major goal in regenerative tendon therapies. Based on the early hypothesis of MSC engraftment and tenogenic differentiation, it was assumed that the differentiated cells would subsequently synthesize new tendon ECM. Indeed, MSCs are capable to synthesize a considerable amount of extracellular matrix even in an undifferentiated state [160]. Furthermore, the composition of the ECM synthesized by differentiated MSC reflects the respective tissue lineage, which is well-established for their chondrogenic or osteogenic differentiation. Corresponding in vitro data exist for the differentiation into the tenogenic lineage, although not always consistent between studies. There is also in vivo evidence that MSC transplantation improves tendon ECM structure. However, this is not necessarily due to ECM synthesis by the MSC themselves but might also be a consequence of protective and stimulatory effects on tenocytes, which in turn might be capable to synthesize the new ECM. Moreover, importantly, there is not simply a lack of ECM in tendinopathy but rather a dysfunctional ECM composition and structure, due to the imbalance of remodeling activities. Particularly, in later stages of the disease, chondroid degeneration and fibrosis impair ECM functionality, thus effective ECM regeneration would also comprise its remodeling and the restoration of physiological remodeling activity within the tendon.

4.1 In vitro evidence

As most tenogenic differentiation studies investigated the expression and/or deposition of tendon-specific extracellular matrix molecules as a marker for successful differentiation, there is quite extensive evidence that the ECM synthesis by MSC is altered during tenogenic differentiation. However, there is some discrepancy between different studies as to whether the ECM molecule expression pattern of tenogenic MSC truly corresponds to that of healthy tendon tissue.

Collagen I, the most abundant protein in healthy tendons, was shown to be upregulated by ectopic mohawk or scleraxis expression [52], in response to treatment with TGF-β superfamily growth factors [60, 67, 88, 93] or scaffold stiffness and alignment [61–63, 74, 81], as well as in three-dimensional dynamic cultures with uniaxial cyclic loading [58, 64, 77, 87]. Furthermore, co-culture with tenocytes in hypoxic conditions or integration of integrin-binding peptides in the scaffold increased collagen I expression on mRNA as well as protein level [69, 72]. However, in other studies, no collagen I upregulation was observed in response to growth factors such as TGF-β [49] or cyclic loading in two-dimensional ASC or BMSC cultures, respectively [66]. Data are particularly conflicting with regard to whether the presence of tendon ECM components promotes or counteracts collagen I expression [46, 47, 58, 64, 65, 83, 84, 88]. Furthermore, even if collagen I is upregulated, which would enable the MSC to contribute to tendon ECM synthesis, this often occurs in conjunction with the upregulation of other extracellular matrix molecules, such as collagen III, decorin, tenascin-C, or cartilage oligomeric matrix protein [60, 61, 69, 70, 72, 74, 77, 83]. While these molecules are important components of native tendon ECM, contributing to collagen organization and fibrillogenesis, their increased presence is also indicative of tendon degeneration or fibrosis [161–163]. Therefore, in order to achieve a beneficial ECM replacement by MSC, their ECM synthesis would have to be highly balanced. It is not yet sufficiently proven that this can be achieved by inducing tenogenic differentiation.

With respect to the hypothesis of active ECM remodeling by MSC, comparatively few data exist so far. Treatment with BMP-12 induced an enhanced secretion of MMP-1 and -8 by ASC [93]. Similarly, ASC culture in collagen scaffolds increased MMP-1, -2, -8, -9, and -13 gene expression and MMP activity compared to two-dimensional culture [46]. For tendon-derived stem cells, it was also found that cyclic mechanical loading did not only upregulate ECM-related genes but also the integrins α1, -α2, and -α11, as well as MMP-9, -13, and -14 [164]. Thus, tenogenic stimuli may increase expression and activation of MMP by MSC. Furthermore, it was found that BMSC inhibits MMP activity in the cell culture medium through secretion of TIMP-1 and TIMP-2, even in an inflammatory environment [165], but that BMSC as well as ASC accumulate active MMP at their cell surface [166]. Although these latter two studies did not focus on tendon therapies, they suggest that MSCs could contribute to matrix remodeling in a highly targeted manner.

Some studies also provide first insight into the interplay of MSC and tenocytes/tendon ECM in matrix remodeling and will therefore be addressed in more detail. In direct co-cultures of ASC and tenocytes, a different temporal regulation of MMP and ECM components was observed compared to tenocytes alone [105]. This included the upregulation of collagen I and tenascin-C gene expression at day 7 and downregulation of tenascin-C and collagen III at later time points (14 and 21 days, respectively) and a higher collagen I to collagen III ratio on protein level at day 7. MMP-1, -2 and -3, as well as TIMP-1 gene expression, increased over time in tenocytes alone but showed a different temporal regulation pattern in the co-cultures with a significantly increased MMP-3 expression at day 7 [105]. A different study from the same group investigated the indirect co-culture of ASC and tendon explants [104]. Here, total protease activity was increased in the co-cultures at day 3, as were the collagenases (putatively MMP-1 and -14) but not the stromelysins MMP-3 and -10. Furthermore, collagen III and tenascin-C deposition by ASC were reduced at day 7. Histology also suggested that ASCs had protective effects on the explant structure, but this was not consistent between donors [104]. However, seemingly in contrast to these findings, MMP-8, -9, and -13 expression by ASC in collagen scaffolds was lower upon stimulation with tendon ECM extract [46], and microvesicles from amniotic membrane mesenchymal cells induced a downregulation of MMP-1, -9, and -13 in tenocytes [101]. Thus, while it can be assumed that MSC actively contribute to and/or modulate tendon ECM remodeling, the exact temporal regulation and context-sensitivity of this mechanism need to be addressed in future studies.

4.2 In vivo evidence

Several in vivo studies have investigated the effect of MSC treatment on tendon ECM composition and structure, as well as on tendon biomechanical parameters. In most of these studies, including an equine large animal study with a follow-up of 45 weeks, the ECM composition was improved by BMSC and ASC treatment, with higher expression of collagen I on gene and/or protein level [106, 114, 120, 122, 140]. Collagen III expression was found to be decreased after ASC implantation [110, 125, 126] but increased after BMSC implantation [106, 122]. Tenascin-C and decorin were found to be increased following BMSC and ASC treatment [112, 114, 140], and glycosaminoglycans were decreased after BMSC treatment [141]. Based on these data, MSCs appear to increase collagen I deposition in healing tendons. Furthermore, as an increase of human-specific collagen I and tenascin-C was demonstrated in a rat model after human ASC implantation, there is also some evidence that MSCs actively contribute to the synthesis of new ECM [114]. The contribution of collagen III, tenascin-C, and decorin synthesis/modulation to tendon healing is to be considered controversially, as illustrated above, and certainly depends on its balance with regard to other ECM components. Yet, beyond mere collagen I synthesis, BMSC and ASC have also repeatedly been shown to improve the structural organization of healing tendons, again including the study with a 45-week follow-up, as well as an experimental trial in horses with naturally occurring tendinopathy [108, 115, 121, 140, 141]. In conjunction with the synthesis and protection of desired ECM components such as collagen I, this could be due to active ECM remodeling and the contribution of synthesized small ECM molecules to collagen fibrillogenesis. Still, it should be acknowledged that some studies in the equine model could demonstrate only few compositional or structural improvements 5 months after ASC treatment [133, 137]. Moreover, despite generally improved ECM structure and collagen I synthesis, collagen II deposits and areas staining positive for alizarin red were found in BMSC-treated tendons [106], suggesting that erroneous MSC differentiation toward the chondrogenic and osteogenic lineage had occurred. Nevertheless, functional testing of BMSC- and ASC-treated tendons indicated an improvement of functional parameters in the majority of studies [107, 108, 112–115, 117, 119, 121, 122], which represents a beneficial effect that can be attributed to ECM regeneration [3].

So far, very few in vivo studies have investigated the effect of MSC on the presence and activation of matrix-remodeling enzymes and their endogenous inhibitors. In the equine model, MMP-13 activity was decreased 6 months after BMSC treatment [141], and MMP-3 gene expression was upregulated in the healing tendons 45 weeks after BMSC treatment [140]. Together, these results might suggest that collagen degradation could be inhibited while degradation of small ECM components is promoted. However, there is much overlap regarding MMP substrates [167], and other studies found no significant differences in MMP and TIMP expression due to ASC treatment [112]. Further studies have to substantiate this hypothesis.

When MSCs were combined with tenogenic growth factors, conflicting results were reported. Treatment with ASC and GDF-5 decreased MMP-2 and TIMP-2 expression and resulted in inferior biomechanical properties compared to ASC treatment alone [112]. Treatment with ASC and BMP-12 promoted ECM degradation, which was interpreted as a side effect of the fibrin-based delivery vehicle [124], but improved tendon ECM regeneration when delivered as cell sheets without fibrin [123]. Interestingly, the latter study showed that this may have been mediated by modulating the ECM remodeling activity of macrophages [123]. A further study from the same group demonstrated beneficial effects of combined ASC and CTGF treatment, although not evaluating effects of ASC alone [125]. A different study showed that predifferentiated BMSC sheets, induced by stepwise stimulation with TGF-β1 and CTGF, resulted in superior tendon regeneration, including improved biomechanical properties than BMSC alone [54]. However, in this study, again, fibrin was used for delivery of noninduced cells, which may have contributed to the differences observed. Thus, although some data suggest that the additional use of growth factors potentiates the beneficial effects of MSC on ECM regeneration, more evidence supporting this hypothesis is required. It should also be acknowledged that growth factor supplementation might impair other regenerative mechanisms of MSC at the same time [93].

5. Immunomodulation

There is a substantial body of evidence that demonstrates the immunomodulatory potential of MSC. While not all underlying mechanisms have been elucidated in detail yet, it is well-understood that MSCs suppress T cell proliferation and promote the modulatory M2 macrophage phenotype [168]. Furthermore, small ECM molecules synthesized by the MSC, such as tenascin-C and decorin, could contribute to immunomodulation [163, 169]. Therefore, it is likely that immunomodulation plays an important role in MSC-based tendon therapies. Against that background, it appears surprising that relatively few studies have addressed the interplay between MSC and the immune system in the context of tendon disease. This may be due to the long-existing perception that inflammation is absent during most stages of tendon disease, which, however, has been changing [5, 170]. While so far existing findings are summarized in the following, immunomodulation in the context of tendon disease will remain a promising field of future research.

5.1 In vitro evidence

In vitro evidence for MSC immunomodulation in tendon disease is scarce. The most comprehensive study investigated whether ASCs influence the effects of differently polarized macrophages on tenocytes in a tri-culture system [98]. In co-cultures of M1 macrophages and tenocytes, release of inflammatory mediators, such as PGE2 and IL-1β, was increased compared to M1 macrophage cultures alone or compared to co-cultures with M0 or M2 macrophages, suggesting inflammatory tenocyte activation. When ASCs were directly co-cultured with the macrophages for 5 days, with the tenocytes added for the last 24 h, tenocyte activation was decreased, with significantly lower release of TNF-α and IL-1β in tri-cultures with M1 macrophages. At the same time, the presence of ASC had increased CD206 expression in M0 and M1 macrophage populations, indicating a switch toward the anti-inflammatory M2 macrophage phenotype and providing insight into the suppressive mechanism. However, ASCs did not effectively counteract inflammatory activation of tenocytes by IL-1β, even when ASCs had been primed with IFN-γ [98].

Interestingly, it has also been shown that tenogenic differentiation of BMSC induced by GDF-5 involves arachidonic acid production and signaling pathways [67], suggesting a link between differentiation and inflammatory processes. In this line, addition of BMP-12 increased IL-6 secretion by ASC and attenuated the suppressive effect of ASC in a mixed lymphocyte reaction [93]. Microvesicles from amniotic membrane mesenchymal cells downregulated TNF-α expression in tenocytes but in contrast to conditioned medium, they had no effect on peripheral blood mononuclear cell proliferation [101, 171]. These studies provide preliminary insight into the modulation of inflammatory tenocyte activation by MSC, while they also suggest that their immunomodulatory potential may be higher when not tenogenically differentiated. Yet, MSC immunomodulation is highly context-specific and influenced by a variety of factors including three-dimensional culture environments as well as inflammatory priming/licensing [172, 173]. Therefore, it remains crucial to perform further studies specifically mimicking aspects of tendon pathophysiology.

5.2 In vivo evidence

The most insightful studies were performed by the same group, shedding light on ASC-mediated immunomodulation in tendon healing in the canine model [123–126]. Corresponding to the group's in vitro findings, ASC alone, delivered via cell sheets, stimulated the anti-inflammatory M2 macrophage phenotype in healing tendons and reduced total mononuclear cell infiltration. The M2 macrophage markers CD163, MRC1, and CD204 were increased on mRNA and/or protein level, as well as IL-4, prostaglandin reductase-1, and VEGF [123, 126]. Combined administration of ASC and BMP-12 promoted these effects, particularly with respect to IL-4 expression [123]. Furthermore, combined treatment with ASC and CTGF decreased IL-1β, IL-6, and IFN-γ and increased IL-4 expression [125]. These latter findings challenge the hypothesis that tenogenic differentiation decreases the MSC immunomodulatory potential. However, when the inflammatory reaction at the tendon repair site was promoted by a fibrin-based delivery vehicle, ASC and BMP-12 further fostered these unwanted effects [124]. This might indicate that strong inflammation alters the MSC immunomodulatory properties toward a proinflammatory phenotype. In contrast, priming with TNF-α increased the anti-inflammatory effects of BMSC: While nonprimed as well as primed BMSC increased IL-10 and reduced IL-1α, primed BMSC also reduced IL-12 and the numbers of M1 macrophages and increased IL-4 and the numbers of M2 macrophages in rat Achilles tendon defects [118]. Further evidence of anti-inflammatory effects of BMSC in tendon healing was demonstrated in a rat model, in which TNF-α, IFN-γ, and IL-1β were reduced, along with an increase of IL-2 and growth factors, including VEGF [111]. Apparently in contrast to most of these findings, however, we observed that clinical signs of inflammation were increased by ASC treatment in the equine model, although this effect was transient [137]. This again illustrates that MSCs can also adopt a pro-inflammatory phenotype and raises questions as to how and whether this should be controlled. When addressing this issue, it should be acknowledged that a certain extent of inflammation is required to drive resolution. In this respect, macrophages and their M2 polarization driven by MSC may play a particularly important role.

6. Trophic support and pro-angiogenetic effects

In addition to the direct effects of MSC on ECM composition and immune cells, trophic support and protection of resident cells are likely to contribute to beneficial effects of MSC in tendon healing. Tenocytes and tendon stem cells rescued by the MSC may be enabled to promote ECM regeneration and counteract inflammation. Furthermore, a MSC-mediated increase in vascularity may be beneficial at least in some stages of tendon healing, as it would improve energy and oxygen supply, as well as disposal of metabolites, thus reduce oxidative and metabolic stress. The presence of vascular endothelial cells, as well as the combination of tenogenic growth factors with VEGF, has also been shown to promote tenogenic differentiation [60, 74]. However, increased vascularity is also associated with tendinopathy pathogenesis and may foster neurogenic inflammation [6], thus this issue is discussed controversially.

Trophic effects on tenocytes were demonstrated in vitro, when ASC and BMSC, as well as BMSC-conditioned medium, promoted the proliferation of tenocytes [94, 102, 103]. Furthermore, ASC as well as BMSC-conditioned medium promoted tenocyte migration [102, 103], and ASC promoted healing in a microwound model [92]. In vivo, results are inconsistent as to whether BMSC and ASC decrease [137, 141] or increase [117] cellularity within healing tendons. However, the rate of apoptosis was lower following BMSC treatment [107], suggesting protective effects of the MSC. Moreover, ASC combined with CTGF locally increased the numbers of CD146-positive tendon stem cells, suggesting an activation and possible rescue of this endogenous cell population [125].

Pro-angiogenetic effects were observed in small, as well as large animal studies, which demonstrated that BMSC and ASC implantation increased vascularity [106, 129, 131], likely mediated by an increase in VEGF (see below). Yet, the opposite effect was observed in horses suffering from naturally occurring tendinopathy following implantation of BMSC [141].

With respect to possible growth factor signaling, in vitro, higher TGF-β bioactivity was found in the BMSC secretome compared to tenocytes [100]. Upon tenogenic differentiation of ASC using BMP-12, VEGF secretion was significantly increased, although no effect on TGF-β was observed [93]. First in vivo evidence regarding the contribution of growth factors in tendon healing following BMSC or ASC implantation was obtained in rat models, in which VEGF, TGF-β, and hepatocyte growth factor expression were increased in the MSC treatment groups [106, 111, 112]. Yet, these studies did not comprehensively reveal whether these factors were released by the MSC or other cells within the tendon lesion.

The brevity of this subsection illustrates that the insight into trophic and protective mechanisms, as well as growth factor release by MSC, in the context of tendon therapies is still limited. Further research is crucial to improve our ability to exploit these effects and, last not least, to prevent potential negative effects associated with some growth factors, such as hypervascularization in response to VEGF or fibrosis in response to TGF-β.

7. Discussion

This review aimed to compile the evidence supporting specific mechanisms of action that may contribute to tendon regeneration in MSC-based cellular therapies. The analysis of the recent literature demonstrated an imbalance between the numbers of studies investigating tenogenic differentiation in vitro and ECM regeneration in vivo and the numbers of studies elucidating other potential mechanisms. This is conceivable as most studies investigating MSC in the context of tendon disease did not specifically aim at clarifying the mechanisms of action. Particularly, the in vivo studies mostly addressed MSC efficacy, at which ECM characteristics are reasonable outcome parameters. Still, despite the overlap with tissue engineering, the overrepresentation of tenogenic differentiation studies may reflect a delay in the field of tendon research. Tendon pathophysiology itself is still not well-understood, making it challenging to transfer the rapidly changing perception of MSC into experimental settings relevant to tendon disease in a timely manner. Yet, it can be anticipated that the general understanding of MSC mechanisms will be successively incorporated into tendon research in the following years.

Taking into account the existing data, the best-evidenced beneficial effect of MSC in tendon regeneration is the improved ECM regeneration. MSCs may also protect and rescue resident tendon cells, but only few data support this hypothesis so far. Both, ECM regeneration and tendon cell protection, are likely to be mediated by a range of mechanisms acting in concert. These may be active over long periods of time, as the engraftment of MSC within tendon lesions was repeatedly demonstrated.

The possible mechanisms mediating ECM regeneration include ECM synthesis and targeted remodeling by the engrafted MSC, inhibition of MMP over-activation, modulation of immune cells with suppression of macrophage-mediated matrix degradation, and modulation of growth factor signaling. Last but not least, the rescue of resident tendon cells could prevent ongoing ECM degeneration, and their trophic support and stimulation by MSC-derived growth factors could re-initiate ECM synthesis and a healthy state of ECM remodeling driven by the tenocytes. A varying extent of evidence supports these different mechanisms, with the collectively most convincing data available for ECM synthesis, immunomodulation, and VEGF-mediated angiogenesis. **Figure 3** illustrates the possible interplay between the different mechanisms and their potential synergies.

The figure summarizes the currently known mechanisms of MSC that may contribute to tendon regeneration. Mechanisms for which there is conclusive evidence from in vivo studies are designated in bold typeface.

However, there may also be antagonisms between different mechanisms, although the evidence is not yet entirely conclusive. Perhaps, tenogenic differentiation and immunomodulation may not occur at the same time. Tenogenic differentiation was shown to interfere with the immunomodulatory potential of MSC [93], and inflammatory environment compromised tenogenic MSC properties [84]. Yet, some in vivo studies revealed anti-inflammatory effects of combined MSC and

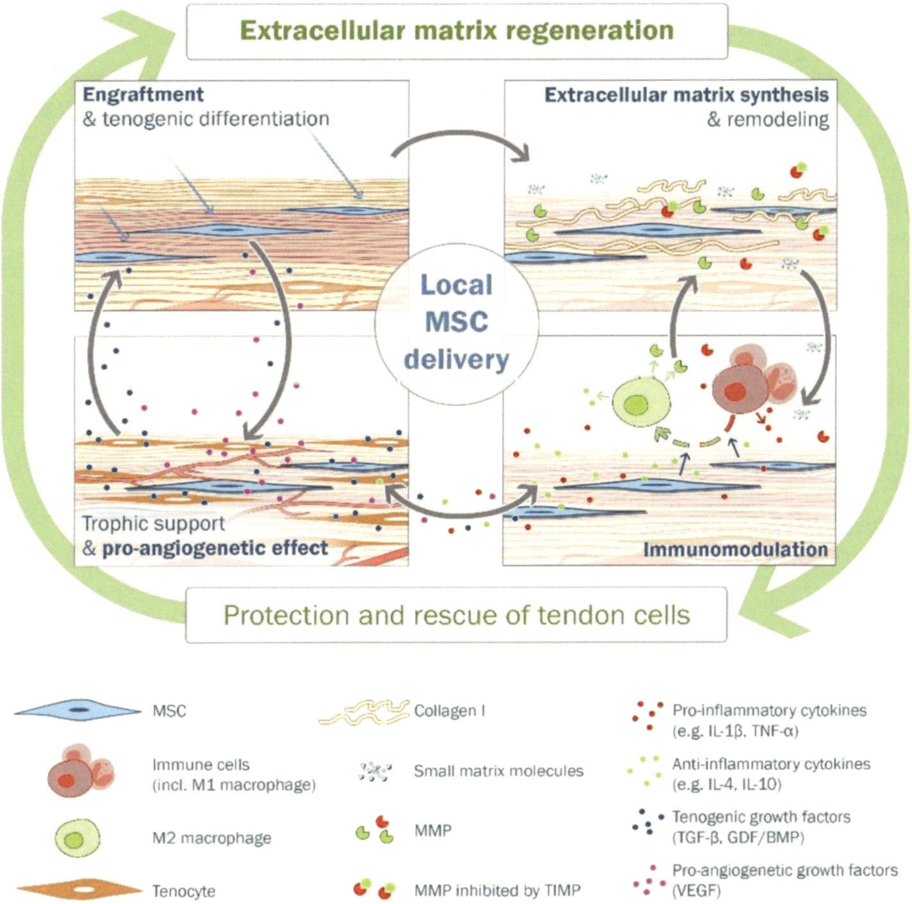

Figure 3.
Mechanisms of action of MSC in tendon healing.

tenogenic growth factor administration [123, 125], although it remained unclear if the MSCs had undergone tenogenic differentiation. It is possible that the context, i.e., the stage of tendon disease, may favor one mechanism over the other. For example, immune cells such as macrophages are not predominating during subclinical stages [6], and the macrophage polarization pattern is distinct in acute vs. chronic disease [8], which will certainly impact on the activation of MSC immunomodulatory mechanisms.

A range of limitations impedes a coherent interpretation of the existing data. These include the different treatment approaches chosen and models used, which make it difficult to elucidate specific reasons for contradictory findings. Inter-donor variability is a further issue that may obscure clarity of findings in studies using human or large animal MSCs [100, 143]. Furthermore, although tenogenic differentiation has extensively been studied, there is neither a consensus on differentiation protocols nor have specific markers for tenogenic differentiation been used consistently. Next, the limited understanding of tendon (patho)physiology makes it difficult to judge whether certain effects observed are beneficial or rather detrimental, e.g., with respect to MMP or TGF-β activity. Last but not least, the illustrated imbalance between evidence levels for particular mechanisms makes it difficult to draw a comprehensive picture at the moment.

8. Conclusion

This review demonstrates progress but also substantial weaknesses which still exist in our understanding of MSC-based cellular tendon therapy and the MSC mechanisms of action in tendon healing. Therefore, considering the low level of clinical evidence, at the moment, MSC-based treatment of tendinopathy appears only justified in the framework of clinical studies. Otherwise, although clinical translation appears temptingly close, it may be wiser to slow down the pace and focus on research into MSC mechanisms in relevant disease models to eventually be able to coax the MSCs toward targeted tendon regeneration.

Acknowledgements

I thank Dr. Wael Kafienah, University of Bristol, UK, for his input during drafting this manuscript, and Nina Schiller, Berlin, Germany, for the graphic design of **Figure 3**. This work was supported by the German Research Foundation (DFG BU3110/1-1).

Conflict of interest

The author has no conflict of interest to declare.

Author details

Janina Burk
Institute of Veterinary Physiology and Saxon Incubator for Clinical Translation, University of Leipzig, Leipzig, Germany

*Address all correspondence to: burk@rz.uni-leipzig.de

References

[1] Docking SI, Rio E, Cook J, Orchard JW, Fortington LV. The prevalence of Achilles and patellar tendon injuries in Australian football players beyond a time-loss definition. Scandinavian Journal of Medicine & Science in Sports. 2018;**28**:2016-2022. DOI: 10.1111/sms.13086

[2] Noback PC, Freibott CE, Tantigate D, Jang E, Greisberg JK, Wong T, et al. Prevalence of asymptomatic Achilles tendinosis. Foot & Ankle International. 2018;**39**:1205-1209. DOI: 10.1177/1071100718778592

[3] Hammer N, Huster D, Fritsch S, Hädrich C, Koch H, Schmidt P, et al. Do cells contribute to tendon and ligament biomechanics? PLoS One. 2014;**9**:e105037. DOI: 10.1371/journal.pone.0105037

[4] Zhang J, JH-C W. The effects of mechanical loading on tendons—An in vivo and in vitro model study. PLoS One. 2013;**8**:e71740. DOI: 10.1371/journal.pone.0071740

[5] Schulze-Tanzil G, Al-Sadi O, Wiegand E, Ertel W, Busch C, Kohl B, et al. The role of pro-inflammatory and immunoregulatory cytokines in tendon healing and rupture: New insights. Scandinavian Journal of Medicine & Science in Sports. 2011;**21**:337-351. DOI: 10.1111/j.1600-0838.2010.01265.x

[6] Abate M, Silbernagel KG, Siljeholm C, Di Iorio A, de Amicis D, Salini V, et al. Pathogenesis of tendinopathies: Inflammation or degeneration? Arthritis Research and Therapy. 2009;**11**:235. DOI: 10.1186/ar2723

[7] Del Buono A, Oliva F, Osti L, Maffulli N. Metalloproteases and tendinopathy. Muscles, Ligaments and Tendons Journal. 2013;**3**:51-57. DOI: 10.11138/mltj/2013.3.1.051

[8] Dakin SG, Werling D, Hibbert A, Abayasekara DR, Young NJ, Smith RK, et al. Macrophage sub-populations and the Lipoxin A(4) receptor implicate active inflammation during equine tendon repair. PLoS One. 2012;**7**:e32333. DOI: 10.1371/journal.pone.0032333

[9] Dakin SG, Buckley CD, Al-Mossawi MH, Hedley R, Martinez FO, Wheway K, et al. Persistent stromal fibroblast activation is present in chronic tendinopathy. Arthritis Research and Therapy. 2017;**19**:16. DOI: 10.1186/s13075-016-1218-4

[10] Hsieh C-F, Yan Z, Schumann RG, Milz S, Pfeifer CG, Schieker M, et al. In vitro comparison of 2d-cell culture and 3d-cell sheets of Scleraxis-programmed bone marrow derived mesenchymal stem cells to primary tendon stem/progenitor cells for tendon repair. International Journal of Molecular Sciences. 2018;**19**:2272. DOI: 10.3390/ijms19082272

[11] Tempfer H, Wagner A, Gehwolf R, Lehner C, Tauber M, Resch H, et al. Perivascular cells of the supraspinatus tendon express both tendon- and stem cell-related markers. Histochemistry and Cell Biology. 2009;**131**:733-741. DOI: 10.1007/s00418-009-0581-5

[12] Bi Y, Ehirchiou D, Kilts TM, Inkson CA, Embree MC, Sonoyama W, et al. Identification of tendon stem/progenitor cells and the role of the extracellular matrix in their niche. Nature Medicine. 2007;**13**:1219-1227

[13] Xu H, Liu F. Downregulation of FOXP1 correlates with tendon stem/progenitor cells aging. Biochemical and Biophysical Research Communications. 2018;**504**:96-102. DOI: 10.1016/j.bbrc.2018.08.136

[14] Gehwolf R, Wagner A, Tempfer H, Tauber M, Bauer H-C. Tendon progenitor cells - their appearance and distribution in degenerated and ageing tendon. Journal of Stem Cells & Regenerative Medicine. 2010;**6**:129

[15] Jensen AR, Kelley BV, Mosich GM, Ariniello A, Eliasberg CD, Vu B, et al. Neer award 2018: Platelet-derived growth factor receptor α co-expression typifies a subset of platelet-derived growth factor receptor β-positive progenitor cells that contribute to fatty degeneration and fibrosis of the murine rotator cuff. Journal of Shoulder and Elbow Surgery. 2018;**27**:1149-1161. DOI: 10.1016/j.jse.2018.02.040

[16] Agarwal S, Loder SJ, Cholok D, Peterson J, Li J, Breuler C, et al. Scleraxis-lineage cells contribute to ectopic bone formation in muscle and tendon. Stem Cells (Dayton, Ohio). 2017;**35**:705-710. DOI: 10.1002/stem.2515

[17] Abat F, Alfredson H, Cucchiarini M, Madry H, Marmotti A, Mouton C, et al. Current trends in tendinopathy: Consensus of the ESSKA basic science committee. Part II: Treatment options. Journal of Experimental Orthopaedics. 2018;**5**:38. DOI: 10.1186/s40634-018-0145-5

[18] Abat F, Alfredson H, Cucchiarini M, Madry H, Marmotti A, Mouton C, et al. Current trends in tendinopathy: Consensus of the ESSKA basic science committee. Part I: Biology, biomechanics, anatomy and an exercise-based approach. Journal of Experimental Orthopaedics. 2017;**4**:18. DOI: 10.1186/s40634-017-0092-6

[19] Dakin SG, Dudhia J, Smith RKW. Science in brief: Resolving tendon inflammation. A new perspective. Equine Veterinary Journal. 2013;**45**:398-400. DOI: 10.1111/evj.12030

[20] Krogh TP, Ellingsen T, Christensen R, Jensen P, Fredberg U. Ultrasound-guided injection therapy of Achilles tendinopathy with platelet-rich plasma or saline: A randomized, blinded, placebo-controlled trial. The American Journal of Sports Medicine. 2016;**44**:1990-1997. DOI: 10.1177/0363546516647958

[21] Di Matteo B, Filardo G, Kon E, Marcacci M. Platelet-rich plasma: Evidence for the treatment of patellar and Achilles tendinopathy—A systematic review. Musculoskeletal Surgery. 2015;**99**:1-9. DOI: 10.1007/s12306-014-0340-1

[22] Li Y, Dai G, Shi L, Lin Y, Chen M, Li G, et al. The potential roles of tendon stem/progenitor cells in tendon ageing. Current Stem Cell Research and Therapy. 2016;**14**(1):34-42. DOI: 10.2174/1574888X13666181017112233

[23] da Silva Meirelles L, Chagastelles PC, Nardi NB. Mesenchymal stem cells reside in virtually all post-natal organs and tissues. Journal of Cell Science. 2006;**119**:2204-2213

[24] Caplan AI. New MSC: MSCs as pericytes are sentinels and gatekeepers. Journal of Orthopaedic Research : Official Publication of the Orthopaedic Research Society. 2017;**35**:1151-1159. DOI: 10.1002/jor.23560

[25] Bourin P, Bunnell BA, Casteilla L, Dominici M, Katz AJ, March KL, et al. Stromal cells from the adipose tissue-derived stromal vascular fraction and culture expanded adipose tissue-derived stromal/stem cells: A joint statement of the International Federation for Adipose Therapeutics and Science (IFATS) and the International Society for Cellular Therapy (ISCT). Cytotherapy. 2013;**15**:641-648. DOI: 10.1016/j.jcyt.2013.02.006

[26] Dominici M, Le Blanc K, Mueller I, Slaper-Cortenbach I, Marini F, Krause D, et al. Minimal criteria for defining multipotent mesenchymal stromal cells. The International Society for Cellular Therapy position statement. Cytotherapy. 2006;**8**:315-317. DOI: 10.1080/14653240600855905

[27] Caplan AI, Bruder SP. Mesenchymal stem cells: Building blocks for molecular medicine in the 21st century. Trends in Molecular Medicine. 2001;7:259-264

[28] Caplan AI. Why are MSCs therapeutic? New data: New insight. The Journal of Pathology. 2009;**217**:318-324. DOI: 10.1002/path.2469

[29] Berebichez-Fridman R, Gómez-García R, Granados-Montiel J, Berebichez-Fastlicht E, Olivos-Meza A, Granados J, et al. The holy grail of Orthopedic surgery: Mesenchymal stem cells-their current uses and potential applications. Stem Cells International. 2017;**2017**:2638305. DOI: 10.1155/2017/2638305

[30] Jeong H, Yim HW, Park H-J, Cho Y, Hong H, Kim NJ, et al. Mesenchymal stem cell therapy for ischemic heart disease: Systematic review and meta-analysis. International Journal of Stem Cells. 2018;**11**:1-12. DOI: 10.15283/ijsc17061

[31] Laroni A, de Rosbo NK, Uccelli A. Mesenchymal stem cells for the treatment of neurological diseases: Immunoregulation beyond neuroprotection. Immunology Letters. 2015;**168**:183-190. DOI: 10.1016/j.imlet.2015.08.007

[32] Fitzsimmons REB, Mazurek MS, Soos A, Simmons CA. Mesenchymal stromal/stem cells in regenerative medicine and tissue engineering. Stem Cells International. 2018;**2018**:8031718. DOI: 10.1155/2018/8031718

[33] Galipeau J, Sensébé L. Mesenchymal stromal cells: Clinical challenges and therapeutic opportunities. Cell Stem Cell. 2018;**22**:824-833. DOI: 10.1016/j.stem.2018.05.004

[34] Squillaro T, Peluso G, Galderisi U. Clinical trials with mesenchymal stem cells: An update. Cell Transplantation. 2016;**25**:829-848. DOI: 10.3727/096368915X689622

[35] Wang L-T, Ting C-H, Yen M-L, Liu K-J, Sytwu H-K, Wu KK, et al. Human mesenchymal stem cells (MSCs) for treatment towards immune- and inflammation-mediated diseases: Review of current clinical trials. Journal of Biomedical Science. 2016;**23**:76. DOI: 10.1186/s12929-016-0289-5

[36] Le Blanc K, Frassoni F, Ball L, Locatelli F, Roelofs H, Lewis I, et al. Mesenchymal stem cells for treatment of steroid-resistant, severe, acute graft-versus-host disease: A phase II study. Lancet. 2008;**371**:1579-1586. DOI: 10.1016/S0140-6736(08)60690-X

[37] Young RG, Butler DL, Weber W, Caplan AI, Gordon SL, Fink DJ. Use of mesenchymal stem cells in a collagen matrix for Achilles tendon repair. Journal of Orthopaedic Research. 1998;**16**:406-413

[38] Smith RK, Korda M, Blunn GW, Goodship AE. Isolation and implantation of autologous equine mesenchymal stem cells from bone marrow into the superficial digital flexor tendon as a potential novel treatment. Equine Veterinary Journal. 2003;**35**:99-102

[39] Godwin EE, Young NJ, Dudhia J, Beamish IC, Smith RK. Implantation of bone marrow-derived mesenchymal stem cells demonstrates improved outcome in horses with overstrain injury of the superficial digital flexor tendon. Equine Veterinary

Journal. 2012;**44**:25-32. DOI: 10.1111/j.2042-3306.2011.00363.x

[40] Pacini S, Spinabella S, Trombi L, Fazzi R, Galimberti S, Dini F, et al. Suspension of bone marrow-derived undifferentiated mesenchymal stromal cells for repair of superficial digital flexor tendon in race horses. Tissue Engineering. 2007;**13**:2949-2955

[41] Smith RK, Webbon PM. Harnessing the stem cell for the treatment of tendon injuries: Heralding a new dawn? Br. The Journal of Sports Medicine. 2005;**39**:582-584

[42] Goldberg AJ, Zaidi R, Brooking D, Kim L, Korda M, Masci L, et al. Autologous stem cells in Achilles tendinopathy (ASCAT): Protocol for a phase IIA, single-centre, proof-of-concept study. BMJ Open. 2018;**8**:e021600. DOI: 10.1136/bmjopen-2018-021600

[43] Kim YS, Sung CH, Chung SH, Kwak SJ, Koh YG. Does an injection of adipose-derived mesenchymal stem cells loaded in fibrin glue influence rotator cuff repair outcomes? A clinical and magnetic resonance imaging study. The American Journal of Sports Medicine. 2017;**45**:2010-2018. DOI: 10.1177/0363546517702863

[44] Jo CH, Chai JW, Jeong EC, Oh S, Kim PS, Yoon JY, et al. Intratendinous injection of autologous adipose tissue-derived mesenchymal stem cells for the treatment of rotator cuff disease: A first-in-human trial. Stem Cells (Dayton, Ohio). 2018;**36**:1441-1450. DOI: 10.1002/stem.2855

[45] Pas HIMFL, Moen MH, Haisma HJ, Winters M. No evidence for the use of stem cell therapy for tendon disorders: A systematic review. British Journal of Sports Medicine. 2017;**51**:996-1002. DOI: 10.1136/bjsports-2016-096794

[46] Yang G, Rothrauff BB, Lin H, Gottardi R, Alexander PG, Tuan RS. Enhancement of tenogenic differentiation of human adipose stem cells by tendon-derived extracellular matrix. Biomaterials. 2013;**34**:9295-9306. DOI: 10.1016/j.biomaterials.2013.08.054

[47] Yang G, Rothrauff BB, Lin H, Yu S, Tuan RS. Tendon-derived extracellular matrix enhances transforming growth factor-β3-induced tenogenic differentiation of human adipose-derived stem cells. Tissue Engineering Parts A. 2017;**23**:166-176. DOI: 10.1089/ten.TEA.2015.0498

[48] Kraus A, Woon C, Raghavan S, Megerle K, Pham H, Chang J. Co-culture of human adipose-derived stem cells with tenocytes increases proliferation and induces differentiation into a tenogenic lineage. Plastic and Reconstructive Surgery. 2013;**132**:754e-766e. DOI: 10.1097/PRS.0b013e3182a48b46

[49] Goncalves AI, Rodrigues MT, Lee SJ, Atala A, Yoo JJ, Reis RL, et al. Understanding the role of growth factors in modulating stem cell tenogenesis. PLoS One. 2013;**8**:e83734. DOI: 10.1371/journal.pone.0083734

[50] Gonçalves AI, Rotherham M, Markides H, Rodrigues MT, Reis RL, Gomes ME, et al. Triggering the activation of Activin a type II receptor in human adipose stem cells towards tenogenic commitment using mechanomagnetic stimulation. Nanomedicine: Nanotechnology, Biology and Medicine. 2018;**14**:1149-1159. DOI: 10.1016/j.nano.2018.02.008

[51] Gonçalves AI, Gershovich PM, Rodrigues MT, Reis RL, Gomes ME. Human adipose tissue-derived tenomodulin positive subpopulation of stem cells: A promising source of tendon progenitor cells. Journal of Tissue

Engineering and Regenerative Medicine. 2018;**12**:762-774. DOI: 10.1002/term.2495

[52] Liu H, Zhang C, Zhu S, Lu P, Zhu T, Gong X, et al. Mohawk promotes the tenogenesis of mesenchymal stem cells through activation of the TGFbeta signaling pathway. Stem Cells (Dayton, Ohio). 2015;**33**:443-455. DOI: 10.1002/stem.1866

[53] Cong XX, Rao XS, Lin JX, Liu XC, Zhang GA, Gao XK, et al. Activation of AKT-mTOR Signaling directs Tenogenesis of mesenchymal stem cells. Stem Cells (Dayton, Ohio). 2018;**36**: 527-539. DOI: 10.1002/stem.2765

[54] Yin Z, Guo J, Wu T-Y, Chen X, Xu L-L, Lin S-E, et al. Stepwise differentiation of mesenchymal stem cells augments tendon-like tissue formation and defect repair In vivo. Stem Cells Translational Medicine. 2016;**5**:1106-1116. DOI: 10.5966/sctm.2015-0215

[55] Nam HY, Pingguan-Murphy B, Amir Abbas A, Mahmood Merican A, Kamarul T. The proliferation and tenogenic differentiation potential of bone marrow-derived mesenchymal stromal cell are influenced by specific uniaxial cyclic tensile loading conditions. Biomechanics and Modeling in Mechanobiology. 2015;**14**:649-663. DOI: 10.1007/s10237-014-0628-y

[56] Stanco D, Vigano M, Perucca Orfei C, Di Giancamillo A, Peretti GM, Lanfranchi L, et al. Multidifferentiation potential of human mesenchymal stem cells from adipose tissue and hamstring tendons for musculoskeletal cell-based therapy. Regenerative Medicine. 2015;**10**:729-743. DOI: 10.2217/rme.14.92

[57] Youngstrom DW, LaDow JE, Barrett JG. Tenogenesis of bone marrow-, adipose-, and tendon-derived stem cells in a dynamic bioreactor. Connective Tissue Research. 2016:1-12. DOI: 10.3109/03008207.2015.1117458

[58] Youngstrom DW, Rajpar I, Kaplan DL, Barrett JG. A bioreactor system for in vitro tendon differentiation and tendon tissue engineering. Journal of Orthopaedic Research. 2015;**33**:911-918. DOI: 10.1002/jor.22848

[59] Iannone M, Ventre M, Formisano L, Casalino L, Patriarca EJ, Netti PA. Nanoengineered surfaces for focal adhesion guidance trigger mesenchymal stem cell self-organization and tenogenesis. Nano Letters. 2015;**15**: 1517-1525. DOI: 10.1021/nl503737k

[60] Bottagisio M, Lopa S, Granata V, Talò G, Bazzocchi C, Moretti M, et al. Different combinations of growth factors for the tenogenic differentiation of bone marrow mesenchymal stem cells in monolayer culture and in fibrin-based three-dimensional constructs. Differentiation; Research in Biological Diversity. 2017;**95**:44-53. DOI: 10.1016/j.diff.2017.03.001

[61] Islam A, Mbimba T, Younesi M, Akkus O. Effects of substrate stiffness on the tenoinduction of human mesenchymal stem cells. Acta Biomaterialia. DOI: 10.1016/j.actbio.2017.05.058

[62] Islam A, Younesi M, Mbimba T, Akkus O. Collagen substrate stiffness anisotropy affects cellular elongation, nuclear shape, and stem cell fate toward anisotropic tissue lineage. Advanced Healthcare Materials. 2016;**5**:2237-2247. DOI: 10.1002/adhm.201600284

[63] Younesi M, Islam A, Kishore V, Anderson JM, Akkus O. Tenogenic induction of human MSCs by anisotropically aligned collagen biotextiles. Advanced Functional Materials. 2014;**24**:5762-5770. DOI: 10.1002/adfm.201400828

[64] Qin T-W, Sun Y-L, Thoreson AR, Steinmann SP, Amadio PC, An K-N, et al. Effect of mechanical stimulation on bone marrow stromal cell-seeded tendon slice constructs: A potential engineered tendon patch for rotator cuff repair. Biomaterials. 2015;**51**: 43-50. DOI: 10.1016/j.biomaterials.2015.01.070

[65] Ning L-J, Zhang Y-J, Zhang Y, Qing Q, Jiang Y-L, Yang J-L, et al. The utilization of decellularized tendon slices to provide an inductive microenvironment for the proliferation and tenogenic differentiation of stem cells. Biomaterials. 2015;**52**:539-550. DOI: 10.1016/j.biomaterials.2015.02.061

[66] Brown JP, Galassi TV, Stoppato M, Schiele NR, Kuo CK. Comparative analysis of mesenchymal stem cell and embryonic tendon progenitor cell response to embryonic tendon biochemical and mechanical factors. Stem Cell Research & Therapy. 2015;**6**:89. DOI: 10.1186/s13287-015-0043-z

[67] Tan S-L, Ahmad TS, Ng W-M, Azlina AA, Azhar MM, Selvaratnam L, et al. Identification of pathways mediating growth differentiation factor5-induced tenogenic differentiation in human bone marrow stromal cells. PLoS One. 2015;**10**:e0140869. DOI: 10.1371/journal.pone.0140869

[68] Vuornos K, Bjorninen M, Talvitie E, Paakinaho K, Kellomaki M, Huhtala H, et al. Human adipose stem cells differentiated on braided polylactide scaffolds is a potential approach for tendon tissue engineering. Tissue Engineering Parts A. 2016;**22**:513-523. DOI: 10.1089/ten.tea.2015.0276

[69] Yu Y, Zhou Y, Cheng T, Lu X, Yu K, Zhou Y, et al. Hypoxia enhances tenocyte differentiation of adipose-derived mesenchymal stem cells by inducing hypoxia-inducible factor-1α in a co-culture system. Cell Proliferation. 2016;**49**:173-184. DOI: 10.1111/cpr.12250

[70] Govoni M, Berardi AC, Muscari C, Campardelli R, Bonafè F, Guarnieri C, et al. An engineered multiphase three-dimensional microenvironment to ensure the controlled delivery of cyclic strain and human growth differentiation factor 5 for the tenogenic commitment of human bone marrow mesenchymal stem cells. Tissue Engineering Parts A. 2017;**23**:811-822. DOI: 10.1089/ten.TEA.2016.0407

[71] Le W, Yao J. The effect of myostatin (GDF-8) on proliferation and tenocyte differentiation of rat bone marrow-derived mesenchymal stem cells. The journal of hand surgery Asian-Pacific Volume. 2017;**22**:200-207. DOI: 10.1142/S0218810417500253

[72] Rehmann MS, Luna JI, Maverakis E, Kloxin AM. Tuning microenvironment modulus and biochemical composition promotes human mesenchymal stem cell tenogenic differentiation. Journal of Biomedical Materials Research. Part A. 2016;**104**:1162-1174. DOI: 10.1002/jbm.a.35650

[73] Laranjeira M, Domingues RMA, Costa-Almeida R, Reis RL, Gomes ME. 3D mimicry of native-tissue-fiber architecture guides tendon-derived cells and adipose stem cells into artificial tendon constructs. Small (Weinheim an der Bergstrasse, Germany). 2017;**13**(31):1700689. DOI: 10.1002/smll.201700689

[74] Wu S, Wang Y, Streubel PN, Duan B. Living nanofiber yarn-based woven biotextiles for tendon tissue engineering using cell tri-culture and mechanical stimulation. Acta Biomaterialia. 2017;**62**:102-115. DOI: 10.1016/j.actbio.2017.08.043

[75] Wu S, Peng H, Li X, Streubel PN, Liu Y, Duan B. Effect of scaffold morphology and cell co-culture on

tenogenic differentiation of HADMSC on centrifugal melt electrospun poly (L-lactic acid) fibrous meshes. Biofabrication. 2017;**9**:44106. DOI: 10.1088/1758-5090/aa8fb8

[76] Rinella L, Marano F, Paletto L, Fraccalvieri M, Annaratone L, Castellano I, et al. Extracorporeal shock waves trigger tenogenic differentiation of human adipose-derived stem cells. Connective Tissue Research. 2018:1-13. DOI: 10.1080/03008207.2018.1424147

[77] Subramanian G, Stasuk A, Elsaadany M, Yildirim-Ayan E. Effect of uniaxial tensile cyclic loading regimes on matrix organization and tenogenic differentiation of adipose-derived stem cells encapsulated within 3D collagen scaffolds. Stem Cells International. 2017;**2017**:6072406. DOI: 10.1155/2017/6072406

[78] Le W, Cheah AE-J, Yao J. Ex-vivo tendon repair augmented with bone marrow derived mesenchymal stem cells stimulated with myostatin for tenogenesis. The journal of hand surgery Asian-Pacific Volume. 2018;**23**:47-57. DOI: 10.1142/S2424835518500066

[79] Morita Y, Yamashita T, Toku T, Ju Y. Optimization of differentiation time of mesenchymal-stem-cell to tenocyte under a cyclic stretching with a microgrooved culture membrane and selected measurement cells. Acta of Bioengineering and Biomechanics. 2018;**20**:3-10

[80] Viganò M, Perucca Orfei C, de Girolamo L, Pearson JR, Ragni E, de Luca P, et al. Housekeeping gene stability in human mesenchymal stem and tendon cells exposed to tenogenic factors. Tissue Engineering. Part C, Methods. 2018;**24**:360-367. DOI: 10.1089/ten.TEC.2017.0518

[81] Zhou K, Feng B, Wang W, Jiang Y, Zhang W, Zhou G, et al. Nanoscaled and microscaled parallel topography promotes tenogenic differentiation of ASC and neotendon formation in vitro. International Journal of Nanomedicine. 2018;**13**:3867-3881. DOI: 10.2147/IJN.S161423

[82] Wang D, Jiang X, Lu A, Tu M, Huang W, Huang P. BMP14 induces tenogenic differentiation of bone marrow mesenchymal stem cells in vitro. Experimental and Therapeutic Medicine. 2018;**16**:1165-1174. DOI: 10.3892/etm.2018.6293

[83] Burk J, Plenge A, Brehm W, Heller S, Pfeiffer B, Kasper C. Induction of tenogenic differentiation mediated by extracellular tendon matrix and short-term cyclic stretching. Stem Cells International. 2016;**2016**:7342379. DOI: 10.1155/2016/7342379

[84] Brandt L, Schubert S, Scheibe P, Brehm W, Franzen J, Gross C, et al. Tenogenic properties of mesenchymal progenitor cells are compromised in an inflammatory environment. International Journal of Molecular Sciences. 2018;**19**(9):2549. DOI: 10.3390/ijms19092549

[85] Marmotti A, Peretti GM, Mattia S, Mangiavini L, de Girolamo L, Viganò M, et al. Pulsed electromagnetic fields improve tenogenic commitment of umbilical cord-derived mesenchymal stem cells: A potential strategy for tendon repair-an in vitro study. Stem Cells International. 2018;**2018**:9048237. DOI: 10.1155/2018/9048237

[86] Zhang B, Luo Q, Deng B, Morita Y, Ju Y, Song G. Construction of tendon replacement tissue based on collagen sponge and mesenchymal stem cells by coupled mechano-chemical induction and evaluation of its tendon repair abilities. Acta Biomaterialia. 2018;**74**:247-259. DOI: 10.1016/j.actbio.2018.04.047

[87] Bosworth LA, Rathbone SR, Bradley RS, Cartmell SH. Dynamic

loading of electrospun yarns guides mesenchymal stem cells towards a tendon lineage. Journal of the Mechanical Behavior of Biomedical Materials. 2014;**39**:175-183. DOI: 10.1016/j.jmbbm.2014.07.009

[88] Roth SP, Schubert S, Scheibe P, Groß C, Brehm W, Burk J. Growth factor-mediated tenogenic induction of multipotent mesenchymal stromal cells is altered by the microenvironment of tendon matrix. Cell Transplantation. 2018;**27**:1434-1450. DOI: 10.1177/0963689718792203

[89] Cai T-Y, Zhu W, Chen X-S, Zhou S-Y, Jia L-S, Sun Y-Q. Fibroblast growth factor 2 induces mesenchymal stem cells to differentiate into tenocytes through the MAPK pathway. Molecular Medicine Reports. 2013;**8**:1323-1328. DOI: 10.3892/mmr.2013.1668

[90] Reed SA, Johnson SE. Expression of scleraxis and tenascin C in equine adipose and umbilical cord blood derived stem cells is dependent upon substrata and FGF supplementation. Cytotechnology. 2014;**66**:27-35. DOI: 10.1007/s10616-012-9533-3

[91] Engebretson B, Mussett ZR, Sikavitsas VI. Tenocytic extract and mechanical stimulation in a tissue-engineered tendon construct increases cellular proliferation and ECM deposition. Biotechnology Journal. DOI: 10.1002/biot.201600595

[92] Veronesi F, Torricelli P, Della Bella E, Pagani S, Fini M. In vitro mutual interaction between tenocytes and adipose-derived mesenchymal stromal cells. Cytotherapy. 2015;**17**:215-223. DOI: 10.1016/j.jcyt.2014.10.006

[93] Zarychta-Wiśniewska W, Burdzinska A, Kulesza A, Gala K, Kaleta B, Zielniok K, et al. Bmp-12 activates tenogenic pathway in human adipose stem cells and affects their immunomodulatory and secretory properties. BMC Cell Biology. 2017;**18**:13. DOI: 10.1186/s12860-017-0129-9

[94] Wu T, Liu Y, Wang B, Sun Y, Lee WYW, Xu J, et al. The use of co-cultured mesenchymal stem cells with tendon-derived stem cells as a better cell source for tendon repair. Tissue Engineering Parts A. DOI: 10.1089/ten.TEA.2016.0248

[95] Taylor SE, Vaughan-Thomas A, Clements DN, Pinchbeck G, Macrory LC, Smith RK, et al. Gene expression markers of tendon fibroblasts in normal and diseased tissue compared to monolayer and three dimensional culture systems. BMC Musculoskeletal Disorders. 2009;**10**:27

[96] Clements LE, Garvican ER, Dudhia J, Smith RKW. Modulation of mesenchymal stem cell genotype and phenotype by extracellular matrix proteins. Connective Tissue Research. 2016;**57**:443-453. DOI: 10.1080/03008207.2016.1215442

[97] Garvican ER, Dudhia J, Alves AL, Clements LE, Plessis FD, Smith RK. Mesenchymal stem cells modulate release of matrix proteins from tendon surfaces in vitro: A potential beneficial therapeutic effect. Regenerative Medicine. 2014;**9**:295-308. DOI: 10.2217/rme.14.7

[98] Manning CN, Martel C, Sakiyama-Elbert SE, Silva MJ, Shah S, Gelberman RH, et al. Adipose-derived mesenchymal stromal cells modulate tendon fibroblast responses to macrophage-induced inflammation in vitro. Stem Cell Research & Therapy. 2015;**6**:74. DOI: 10.1186/s13287-015-0059-4

[99] Veronesi F, Della Bella E, Torricelli P, Pagani S, Fini M. Effect of adipose-derived mesenchymal stromal cells on tendon healing in aging and estrogen deficiency: An in vitro co-culture

model. Cytotherapy. 2015;**17**:1536-1544. DOI: 10.1016/j.jcyt.2015.07.007

[100] Ekwueme EC, Shah JV, Mohiuddin M, Ghebes CA, Crispim JF, Saris DBF, et al. Cross-talk between human tenocytes and bone marrow stromal cells potentiates extracellular matrix Remodeling In vitro. Journal of Cellular Biochemistry. 2016;**117**:684-693. DOI: 10.1002/jcb.25353

[101] Lange-Consiglio A, Perrini C, Tasquier R, Deregibus MC, Camussi G, Pascucci L, et al. Equine amniotic microvesicles and their anti-inflammatory potential in a tenocyte model in vitro. Stem Cells and Development. 2016;**25**:610-621. DOI: 10.1089/scd.2015.0348

[102] Long C, Wang Z, Legrand A, Chattopadhyay A, Chang J, Fox PM. Tendon tissue engineering: Mechanism and effects of human tenocyte coculture with adipose-derived stem cells. The Journal of Hand Surgery. 2018;**43**:183.e1-183.e9. DOI: 10.1016/j.jhsa.2017.07.031

[103] Chen Q, Liang Q, Zhuang W, Zhou J, Zhang B, Xu P, et al. Tenocyte proliferation and migration promoted by rat bone marrow mesenchymal stem cell-derived conditioned medium. Biotechnology Letters. 2018;**40**:215-224. DOI: 10.1007/s10529-017-2446-7

[104] Costa-Almeida R, Berdecka D, Rodrigues MT, Reis RL, Gomes ME. Tendon explant cultures to study the communication between adipose stem cells and native tendon niche. Journal of Cellular Biochemistry. 2018;**119**:3653-3662. DOI: 10.1002/jcb.26573

[105] Costa-Almeida R, Calejo I, Reis RL, Gomes ME. Crosstalk between adipose stem cells and tendon cells reveals a temporal regulation of tenogenesis by matrix deposition and remodeling. Journal of Cellular Physiology. 2018;**233**:5383-5395. DOI: 10.1002/jcp.26363

[106] Machova Urdzikova L, Sedlacek R, Suchy T, Amemori T, Ruzicka J, Lesny P, et al. Human multipotent mesenchymal stem cells improve healing after collagenase tendon injury in the rat. Biomedical Engineering Online. 2014;**13**:42. DOI: 10.1186/1475-925X-13-42

[107] Selek O, Buluç L, Muezzinoğlu B, Ergün RE, Ayhan S, Karaöz E. Mesenchymal stem cell application improves tendon healing via anti-apoptotic effect (animal study). Acta Orthopaedica et Traumatologica Turcica. 2014;**48**:187-195. DOI: 10.3944/AOTT.2014.2985

[108] Chen H-S, Su Y-T, Chan T-M, Su Y-J, Syu W-S, Harn H-J, et al. Human adipose-derived stem cells accelerate the restoration of tensile strength of tendon and alleviate the progression of rotator cuff injury in a rat model. Cell Transplantation. 2015;**24**:509-520. DOI: 10.3727/096368915X686968

[109] Al-Ani MK, Xu K, Sun Y, Pan L, Xu Z, Yang L. Study of bone marrow mesenchymal and tendon-derived stem cells transplantation on the regenerating effect of achilles tendon ruptures in rats. Stem Cells International. 2015;**2015**:984146. DOI: 10.1155/2015/984146

[110] Oshita T, Tobita M, Tajima S, Mizuno H. Adipose-derived stem cells improve collagenase-induced tendinopathy in a rat model. The American Journal of Sports Medicine. 2016;**44**:1983-1989. DOI: 10.1177/0363546516640750

[111] Yuksel S, Guleç MA, Gultekin MZ, Adanır O, Caglar A, Beytemur O, et al. Comparison of the early period effects of bone marrow-derived mesenchymal stem cells and platelet-rich plasma on the Achilles tendon

ruptures in rats. Connective Tissue Research. 2016;**57**:360-373. DOI: 10.1080/03008207.2016.1189909

[112] de Aro AA, Carneiro GD, Teodoro LFR, da Veiga FC, Ferrucci DL, Simões GF, et al. Injured achilles tendons treated with adipose-derived stem cells transplantation and GDF-5. Cell. DOI: 10.3390/cells7090127

[113] Adams SB Jr, Thorpe MA, Parks BG, Aghazarian G, Allen E, Schon LC. Stem cell-bearing suture improves Achilles tendon healing in a rat model. Foot & Ankle International. 2014;**35**:293-299. DOI: 10.1177/1071100713519078

[114] Lee SY, Kwon B, Lee K, Son YH, Chung SG. Therapeutic mechanisms of human adipose-derived mesenchymal stem cells in a rat tendon injury model. The American Journal of Sports Medicine. 2017;**45**:1429-1439. DOI: 10.1177/0363546517689874

[115] Peach MS, Ramos DM, James R, Morozowich NL, Mazzocca AD, Doty SB, et al. Engineered stem cell niche matrices for rotator cuff tendon regenerative engineering. PLoS One. 2017;**12**:e0174789. DOI: 10.1371/journal.pone.0174789

[116] Otabe K, Nakahara H, Hasegawa A, Matsukawa T, Ayabe F, Onizuka N, et al. Transcription factor Mohawk controls tenogenic differentiation of bone marrow mesenchymal stem cells in vitro and in vivo. Journal of Orthopaedic Research : Official Publication of the Orthopaedic Research Society. 2015;**33**:1-8. DOI: 10.1002/jor.22750

[117] Chiou GJ, Crowe C, McGoldrick R, Hui K, Pham H, Chang J. Optimization of an injectable tendon hydrogel: The effects of platelet-rich plasma and adipose-derived stem cells on tendon healing in vivo. Tissue Engineering Parts A. 2015;**21**:1579-1586. DOI: 10.1089/ten.TEA.2014.0490

[118] Aktas E, Chamberlain CS, Saether EE, Duenwald-Kuehl SE, Kondratko-Mittnacht J, Stitgen M, et al. Immune modulation with primed mesenchymal stem cells delivered via biodegradable scaffold to repair an Achilles tendon segmental defect. Journal of Orthopaedic Research : Official Publication of the Orthopaedic Research Society. 2017;**35**:269-280. DOI: 10.1002/jor.23258

[119] Behfar M, Javanmardi S, Sarrafzadeh-Rezaei F. Comparative study on functional effects of allotransplantation of bone marrow stromal cells and adipose derived stromal vascular fraction on tendon repair: A biomechanical study in rabbits. Cell Journal. 2014;**16**:263-270

[120] He M, Gan AWT, Lim AYT, Goh JCH, Hui JHP, Chong AKS. Bone marrow derived mesenchymal stem cell augmentation of rabbit flexor tendon healing. Hand Surgery: An International Journal Devoted to Hand and upper limb surgery and related research : journal of the Asia-Pacific Federation of Societies for Surgery of the Hand. 2015;**20**:421-429. DOI: 10.1142/S0218810415500343

[121] Deng D, Wang W, Wang B, Zhang P, Zhou G, Zhang WJ, et al. Repair of Achilles tendon defect with autologous ASCs engineered tendon in a rabbit model. Biomaterials. 2014;**35**:8801-8809. DOI: 10.1016/j.biomaterials.2014.06.058

[122] Cai J, Yang Y, Ai C, Jin W, Sheng D, Chen J, et al. Bone marrow stem cells-seeded polyethylene terephthalate scaffold in repair and regeneration of rabbit achilles tendon. Artificial Organs. DOI: 10.1111/aor.13298

[123] Gelberman RH, Linderman SW, Jayaram R, Dikina AD, Sakiyama-Elbert S, Alsberg E, et al. Combined administration of ASCs and BMP-12 promotes an m2 macrophage phenotype and enhances tendon healing. Clinical

Orthopaedics and Related Research. 2017;**475**:2318-2331. DOI: 10.1007/s11999-017-5369-7

[124] Gelberman RH, Shen H, Kormpakis I, Rothrauff B, Yang G, Tuan RS, et al. Effect of adipose-derived stromal cells and BMP12 on intrasynovial tendon repair: A biomechanical, biochemical, and proteomics study. Journal of Orthopaedic Research : Official Publication of the Orthopaedic Research Society. 2016;**34**:630-640. DOI: 10.1002/jor.23064

[125] Shen H, Jayaram R, Yoneda S, Linderman SW, Sakiyama-Elbert SE, Xia Y, et al. The effect of adipose-derived stem cell sheets and CTGF on early flexor tendon healing in a canine model. Scientific Reports. 2018;**8**:11078. DOI: 10.1038/s41598-018-29474-8

[126] Shen H, Kormpakis I, Havlioglu N, Linderman SW, Sakiyama-Elbert SE, Erickson IE, et al. The effect of mesenchymal stromal cell sheets on the inflammatory stage of flexor tendon healing. Stem Cell Research & Therapy. 2016;**7**:144. DOI: 10.1186/s13287-016-0406-0

[127] Khan MR, Dudhia J, David FH, de Godoy R, Mehra V, Hughes G, et al. Bone marrow mesenchymal stem cells do not enhance intra-synovial tendon healing despite engraftment and homing to niches within the synovium. Stem Cell Research & Therapy. 2018;**9**:169. DOI: 10.1186/s13287-018-0900-7

[128] Scharf A, Holmes S, Thoresen M, Mumaw J, Stumpf A, Peroni J. Superparamagnetic iron oxide nanoparticles as a means to track mesenchymal stem cells in a large animal model of tendon injury. Contrast Media & Molecular Imaging. 2015;**10**:388-397. DOI: 10.1002/cmmi.1642

[129] Lacitignola L, Staffieri F, Rossi G, Francioso E, Crovace A. Survival of bone marrow mesenchymal stem cells labelled with red fluorescent protein in an ovine model of collagenase-induced tendinitis. Veterinary and Comparative Orthopaedics and Traumatology. 2014;**27**:204-209. DOI: 10.3415/VCOT-13-09-0113

[130] Brandão JS, Alvarenga ML, Pfeifer JPH, Dos Santos VH, Fonseca-Alves CE, Rodrigues M, et al. Allogeneic mesenchymal stem cell transplantation in healthy equine superficial digital flexor tendon: A study of the local inflammatory response. Research in Veterinary Science. 2018;**118**:423-430. DOI: 10.1016/j.rvsc.2018.03.012

[131] Conze P, van Schie HT, van WR, Staszyk C, Conrad S, Skutella T, et al. Effect of autologous adipose tissue-derived mesenchymal stem cells on neovascularization of artificial equine tendon lesions. Regenerative Medicine. 2014;**9**:743-757. DOI: 10.2217/rme.14.55

[132] Geburek F, Mundle K, Conrad S, Hellige M, Walliser U, van Schie HT, et al. Tracking of autologous adipose tissue-derived mesenchymal stromal cells with in vivo magnetic resonance imaging and histology after intralesional treatment of artificial equine tendon lesions–A pilot study. Stem Cell Research & Therapy. 2016;**7**:21. DOI: 10.1186/s13287-016-0281-8

[133] Geburek F, Roggel F, van Schie HTM, Beineke A, Estrada R, Weber K, et al. Effect of single intralesional treatment of surgically induced equine superficial digital flexor tendon core lesions with adipose-derived mesenchymal stromal cells: A controlled experimental trial. Stem Cell Research & Therapy. 2017;**8**:129. DOI: 10.1186/s13287-017-0564-8

[134] Becerra P, Valdes Vazquez MA, Dudhia J, Fiske-Jackson AR, Neves F, Hartman NG, et al. Distribution of injected technetium(99m)-labeled mesenchymal stem cells in horses with

naturally occurring tendinopathy. Journal of Orthopaedic Research. 2013;**31**:1096-1102. DOI: 10.1002/jor.22338

[135] Sole A, Spriet M, Padgett KA, Vaughan B, Galuppo LD, Borjesson DL, et al. Distribution and persistence of technetium-99 hexamethyl propylene amine oxime-labelled bone marrow-derived mesenchymal stem cells in experimentally induced tendon lesions after intratendinous injection and regional perfusion of the equine distal limb. Equine Veterinary Journal. 2013;**45**:726-731. DOI: 10.1111/evj.12063

[136] Carvalho AM, Yamada AL, Golim MA, Alvarez LE, Hussni CA, Alves AL. Evaluation of mesenchymal stem cell migration after equine tendonitis therapy. Equine Veterinary Journal. 2014;**46**:635-638. DOI: 10.1111/evj.12173

[137] Ahrberg AB, Horstmeier C, Berner D, Brehm W, Gittel C, Hillmann A, et al. Effects of mesenchymal stromal cells versus serum on tendon healing in a controlled experimental trial in an equine model. BMC Musculoskeletal Disorders. 2018;**19**:230. DOI: 10.1186/s12891-018-2163-y

[138] Berner D, Brehm W, Gerlach K, Gittel C, Offhaus J, Paebst F, et al. Longitudinal cell tracking and simultaneous monitoring of tissue regeneration after cell treatment of natural tendon disease by low-field magnetic resonance imaging. Stem Cells International. 2016;**2016**:1207190. DOI: 10.1155/2016/1207190

[139] Burk J, Berner D, Brehm W, Hillmann A, Horstmeier C, Josten C, et al. Long-term cell tracking following local injection of mesenchymal stromal cells in the equine model of induced tendon disease. Cell Transplantation. 2016;**25**:2199-2211. DOI: 10.3727/096368916X692104

[140] Romero A, Barrachina L, Ranera B, Remacha AR, Moreno B, de Blas I, et al. Comparison of autologous bone marrow and adipose tissue derived mesenchymal stem cells, and platelet rich plasma, for treating surgically induced lesions of the equine superficial digital flexor tendon. Veterinary Journal (London, England : 1997). 2017;**224**:76-84. DOI: 10.1016/j.tvjl.2017.04.005

[141] Smith RK, Werling NJ, Dakin SG, Alam R, Goodship AE, Dudhia J. Beneficial effects of autologous bone marrow-derived mesenchymal stem cells in naturally occurring tendinopathy. PLoS One. 2013;**8**:e75697. DOI: 10.1371/journal.pone.0075697

[142] Le Blanc K. Immunomodulatory effects of fetal and adult mesenchymal stem cells. Cytotherapy. 2003;**5**:485-489. DOI: 10.1080/14653240310003611

[143] Peffers MJ, Collins J, Loughlin J, Proctor C, Clegg PD. A proteomic analysis of chondrogenic, osteogenic and tenogenic constructs from ageing mesenchymal stem cells. Stem Cell Research & Therapy. 2016;**7**:133. DOI: 10.1186/s13287-016-0384-2

[144] Hillmann A, Ahrberg AB, Brehm W, Heller S, Josten C, Paebst F, et al. Comparative characterization of human and equine mesenchymal stromal cells: A basis for translational studies in the equine model. Cell Transplantation. 2016;**25**(1):109-124. DOI: 10.3727/096368915X687822

[145] Phelps J, Sanati-Nezhad A, Ungrin M, Duncan NA, Sen A. Bioprocessing of mesenchymal stem cells and their derivatives: Toward cell-free therapeutics. Stem Cells International. 2018;**2018**:9415367. DOI: 10.1155/2018/9415367

[146] Liu Y, Suen C-W, Zhang J-F, Li G. Current concepts on tenogenic differentiation and clinical applications. Journal of Orthopaedic Translation. 2017;**9**:28-42. DOI: 10.1016/j.jot.2017.02.005

[147] Zhang Y-J, Chen X, Li G, Chan K-M, Heng BC, Yin Z, et al. Concise review: Stem cell fate guided by bioactive molecules for tendon regeneration. Stem Cells Translational Medicine. 2018;**7**:404-414. DOI: 10.1002/sctm.17-0206

[148] Kuo CK, Tuan RS. Mechanoactive tenogenic differentiation of human mesenchymal stem cells. Tissue Engineering. Part A. 2008;**14**:1615-1627

[149] Butler DL, Juncosa-Melvin N, Boivin GP, Galloway MT, Shearn JT, Gooch C, et al. Functional tissue engineering for tendon repair: A multidisciplinary strategy using mesenchymal stem cells, bioscaffolds, and mechanical stimulation. Journal of Orthopaedic Research. 2008;**26**:1-9

[150] Nirmalanandhan VS, Dressler MR, Shearn JT, Juncosa-Melvin N, Rao M, Gooch C, et al. Mechanical stimulation of tissue engineered tendon constructs: Effect of scaffold materials. Journal of Biomechanical Engineering. 2007;**129**:919-923

[151] Agata H, Watanabe N, Ishii Y, Kubo N, Ohshima S, Yamazaki M, et al. Feasibility and efficacy of bone tissue engineering using human bone marrow stromal cells cultivated in serum-free conditions. Biochemical and Biophysical Research Communications. 2009;**382**:353-358. DOI: 10.1016/j.bbrc.2009.03.023

[152] Alberton P, Popov C, Pragert M, Kohler J, Shukunami C, Schieker M, et al. Conversion of human bone marrow-derived mesenchymal stem cells into tendon progenitor cells by ectopic expression of scleraxis. Stem Cells and Development. 2012;**21**:846-858. DOI: 10.1089/scd.2011.0150

[153] Chen YJ, Huang CH, Lee IC, Lee YT, Chen MH, Young TH. Effects of cyclic mechanical stretching on the mRNA expression of tendon/ligament-related and osteoblast-specific genes in human mesenchymal stem cells. Connective Tissue Research. 2008;**49**:7-14

[154] Xu B, Song G, Ju Y, Li X, Song Y, Watanabe S. RhoA/ROCK, cytoskeletal dynamics, and focal adhesion kinase are required for mechanical stretch-induced tenogenic differentiation of human mesenchymal stem cells. Journal of Cellular Physiology. 2012;**227**:2722-2729. DOI: 10.1002/jcp.23016

[155] Maharam E, Yaport M, Villanueva NL, Akinyibi T, Laudier D, He Z, et al. Rho/Rock signal transduction pathway is required for MSC tenogenic differentiation. Bone Research. 2015;**3**:15015. DOI: 10.1038/boneres.2015.15

[156] Chen S, Deng G, Li K, Zheng H, Wang G, Yu B, et al. Interleukin-6 promotes proliferation but inhibits tenogenic differentiation via the Janus kinase/signal transducers and activators of transcription 3 (JAK/STAT3) pathway in tendon-derived stem cells. Medical Science Monitor: International Medical Journal of experimental and Clinical Research. 2018;**24**:1567-1573

[157] Zhang K, Asai S, Yu B, Enomoto-Iwamoto M. IL-1beta irreversibly inhibits tenogenic differentiation and alters metabolism in injured tendon-derived progenitor cells in vitro. Biochemical and Biophysical Research Communications. 2015;**463**:667-672. DOI: 10.1016/j.bbrc.2015.05.122

[158] Liu C, Luo J-W, Liang T, Lin L-X, Luo Z-P, Zhuang Y-Q, et al. Matrix stiffness regulates the differentiation of tendon-derived stem cells through FAK-ERK1/2 activation. Experimental Cell Research. 2018;**373**(1-2):62-70. DOI: 10.1016/j.yexcr.2018.08.023

[159] Safi E, Ficklscherer A, Bondarava M, Betz O, Zhang A, Jansson V, et al. Migration of mesenchymal stem cells

of bursal tissue after rotator cuff repair in rats. Joints. 2018;**6**:4-9. DOI: 10.1055/s-0038-1636948

[160] Anasiz Y, Ozgul RK, Uckan-Cetinkaya D. A new chapter for mesenchymal stem cells: Decellularized extracellular matrices. Stem Cell Reviews. 2017;**13**:587-597. DOI: 10.1007/s12015-017-9757-x

[161] Riley GP, Harrall RL, Cawston TE, Hazleman BL, Mackie EJ. Tenascin-C and human tendon degeneration. The American Journal of Pathology. 1996;**149**:933-943

[162] Dunkman AA, Buckley MR, Mienaltowski MJ, Adams SM, Thomas SJ, Satchell L, et al. Decorin expression is important for age-related changes in tendon structure and mechanical properties. Matrix Biology. 2013;**32**:3-13. DOI: 10.1016/j.matbio.2012.11.005

[163] Yoon JH, Halper J. Tendon proteoglycans: Biochemistry and function. Journal of Musculoskeletal & Neuronal Interactions. 2005;**5**:22-34

[164] Popov C, Burggraf M, Kreja L, Ignatius A, Schieker M, Docheva D. Mechanical stimulation of human tendon stem/progenitor cells results in upregulation of matrix proteins, integrins and MMPs, and activation of p38 and ERK1/2 kinases. BMC Molecular Biology. 2015;**16**:6. DOI: 10.1186/s12867-015-0036-6

[165] Lozito TP, Tuan RS. Mesenchymal stem cells inhibit both endogenous and exogenous MMPs via secreted TIMPs. Journal of Cellular Physiology. 2011;**226**:385-396. DOI: 10.1002/jcp.22344

[166] Lozito TP, Jackson WM, Nesti LJ, Tuan RS. Human mesenchymal stem cells generate a distinct pericellular zone of MMP activities via binding of MMPs and secretion of high levels of TIMPs. Matrix Biology: Journal of the International Society for Matrix Biology. 2014;**34**:132-143. DOI: 10.1016/j.matbio.2013.10.003

[167] Eckhard U, Huesgen PF, Schilling O, Bellac CL, Butler GS, Cox JH, et al. Active site specificity profiling of the matrix metalloproteinase family: Proteomic identification of 4300 cleavage sites by nine MMPs explored with structural and synthetic peptide cleavage analyses. Matrix Biology: Journal of the International Society for Matrix Biology. 2016;**49**:37-60. DOI: 10.1016/j.matbio.2015.09.003

[168] Wang M, Yuan Q, Xie L. Mesenchymal stem cell-based immunomodulation: Properties and clinical application. Stem Cells International. 2018;**2018**:3057624. DOI: 10.1155/2018/3057624

[169] Midwood KS, Chiquet M, Tucker RP, Orend G. Tenascin-C at a glance. Journal of Cell Science. 2016;**129**: 4321-4327. DOI: 10.1242/jcs.190546

[170] Tang C, Chen Y, Huang J, Zhao K, Chen X, Yin Z, et al. The roles of inflammatory mediators and immunocytes in tendinopathy. Journal of Orthopaedic Translation. 2018;**14**: 23-33. DOI: 10.1016/j.jot.2018.03.003

[171] Lange-Consiglio A, Rossi D, Tassan S, Perego R, Cremonesi F, Parolini O. Conditioned medium from horse amniotic membrane-derived multipotent progenitor cells: Immunomodulatory activity in vitro and first clinical application in tendon and ligament injuries in vivo. Stem Cells and Development. 2013;**22**:3015-3024. DOI: 10.1089/scd.2013.0214

[172] Vallés G, Bensiamar F, Crespo L, Arruebo M, Vilaboa N, Saldaña L. Topographical cues regulate the crosstalk between MSCs and macrophages. Biomaterials. 2015;**37**:124-133. DOI: 10.1016/j.biomaterials.2014.10.028

[173] Cassano JM, Schnabel LV, Goodale MB, Fortier LA. Inflammatory licensed equine MSCs are chondroprotective and exhibit enhanced immunomodulation in an inflammatory environment. Stem Cell Research & Therapy. 2018;**9**:82. DOI: 10.1186/s13287-018-0840-2

Chapter 7

Physiology of Flexor Tendon Healing and Rationale for Treatment Protocols

Justin Yousef

Abstract

The hand functions as both a vital tactile organ and a grasping mechanism that is tempered by finely controlled accuracy. Essential to these functions are the delicate movements of the extrinsic flexor tendons. Repair of the injured flexor tendon in the hand to achieve normal function remains a difficult task, and controversy exists as to what postoperative rehabilitation protocols should be utilised. This chapter will focus on the pathophysiology of repair when flexor tendons are ruptured, the unique anatomy of flexor tendons, the latest molecular updates, repair principles, initial management procedures and the rationale for the various postoperative rehabilitation protocols that are used.

Keywords: flexor tendon injury, rehabilitation, tendon healing, molecular updates, complications

1. Introduction

An accurate understanding of the physiology of flexor tendon healing is essential to maximising patient outcomes and justifying the current treatment regimens of surgical repair and postoperative rehabilitation protocols. Our current understanding of flexor tendon healing is a continually evolving area. Therefore, this chapter aims to instruct the reader of the current understanding of flexor tendon basic science, the latest molecular updates, justifications for various surgical and rehabilitation regimens and future research trends. The reader is strongly encouraged to seek alternative resources for more detail regarding flexor tendon anatomy as this will be covered briefly in this chapter. Secondary flexor tendon reconstruction will not be discussed.

2. Macroscopic flexor tendon anatomy

This section will focus only on the flexor sheath and vascular supply. The reader is strongly encouraged to seek the vast array of anatomy texts to familiarise themselves with flexor tendon anatomy, paying attention to the:

- Flexor digitorum superficialis (FDS)

- Flexor digitorum profundus (FDP)
- Flexor pollicis longus (FPL)
- Flexor sheath and pulley system
- Vascular supply

2.1 Flexor sheath and pulleys

The extrinsic flexor tendons of the hand possess true fibro-osseous tunnels in the digits, called the "flexor sheath". Their purpose is to provide very efficient lubrication in an area subject to a change of direction and increase in friction [1]. Proximal to the metacarpophalangeal (MCP) joints, the flexor tendons enter the flexor sheath. This tunnel functions to hold the tendons in close proximity to the phalanges to prevent "bowstringing" and to increase the efficiency of tendon glide [2].

Condensations in the sheath are called pulleys—these almost encircle the flexor tendons to form a fibro-osseous channel that keeps the tendons adjacent to the phalanges [3]. In effect, the pulleys enable the transfer of a translational force generated from the muscle tendon unit into a rotational moment on the phalanges [3]. Pulleys are classified on the basis of their shape—annular or cruciate. There are five annular pulleys (named A1–A5 from proximal to distal) and three cruciate pulleys (named C1–C3 from proximal to distal). The A2 and A4 pulleys insert directly onto the bone over the proximal and middle phalanges, respectively [3]. Traditionally, A2 and A4 are considered to be the pulleys that prevent bowstringing. However, it has now been shown that partial distal excisions of 25% of the A2 pulley, up to 75% of the A4 pulley and 25% of combined A2 and A4 have no significant effect on digit range of motion or work of flexion [4, 5]. The A1, A3 and A5 pulleys are located over the MCP, proximal interphalangeal (PIP) and distal interphalangeal (DIP) joints, respectively [3]. Proximal to the A1 pulley is the palmar aponeurosis (PA) pulley which has been implicated in the aetiology of trigger finger [6].

The cruciate pulleys lie between the A2–A3, the A3–A4 and the A4–A5 pulleys, respectively [3]. These pulleys function like accordions, allowing the sheath to expand and compress with flexion and extension.

Traditionally, it was once thought that the thumb had three pulleys—A1, oblique and A2 [3]. The A1 pulley lies over the MCP joint, the oblique pulley runs from proximal ulnar to distal radial over the proximal phalanx and the A2 pulley is located over the interphalangeal (IP) joint [3]. A fourth pulley, the variable annular pulley, was first reported in 2012 where it was found to be present in 93% of cadaver specimens [7]. It can have three orientations—transverse, oblique or continuous [7].

2.2 Vascular supply of the extrinsic hand flexors

Both FDP and FDS tendons in the digits receive dual nutritional supply from vascular perfusion and synovial diffusion [8]. There is some variation in the vascular system, but, generally speaking, the flexor tendons receive their blood supply via two vincula each—a short and long vinculum [9]. Vincula are folds of mesotenon carrying blood to the tendons [9]:

- The vinculum longus superficialis (VLS) arises from the radial or ulnar side of the base of the proximal phalanx. It receives its blood supply from the transverse communicating branch of the digital arteries at the base of the proximal phalanx.

- The vinculum brevis superficialis (VBS) arises from the volar plate of the PIP joint and attached to the decussation of the FDS. The source vessels are from the proximal transverse digital artery.

- The vinculum brevis profundus (VBP) arises along the distal two thirds of the middle phalanx (whose source arteries are from both the interphalangeal and distal transverse digital arteries) to insert dorsally on the FDP.

- The vinculum longus profundus generally originates at the level of insertion of the FDS tendon (through the decussation of the FDS) and attaches to the FDP directly, and its source is the proximal transverse digital artery.

3. Microscopic aspects of flexor tendon anatomy

3.1 Collagen

Tendons consist of mostly Type I collagen and elastin embedded in a proteoglycan-water matrix [1]. Collagen contributes 65–85% of the dry mass of the tendon [1]. The collagen, elastin and proteoglycan-water matrix are formed by tenoblasts and tenocytes (refer to Section 3.4). These cells are elongated fibroblasts and fibrocytes which lie between the collagen fibres and are organised in a complex hierarchal scheme to form the tendon proper [10]. Soluble tropocollagen molecules form cross-links to create insoluble collagen molecules, which then aggregate progressively into microfibrils and then into visible units under the electron microscope called collagen fibrils [1].

3.2 Collagen fibres, fibre bundles and fascicles

The collagen fibrils in turn aggregate together to form the basic tendon unit—the collagen fibre. The collagen fibre is defined as the smallest tendon unit visible using light microscopy [1]. Aggregates of collagen fibres form a primary fibre bundle called a subfascicle, and a group of primary fibre bundles form a secondary fibre bundle called a fascicle. A group of secondary fascicles in turn form a tertiary bundle; it is the tertiary bundles that contribute to the full tendon and are surrounded by epitenon (refer to Section 3.3).

Both the fascicles and tertiary tendon bundles show a spiral formation along the course of the tendon [1]. In the resting state, the collagen fibres and fibrils show a wavy configuration that appears as regular bands across the fibre surface [11]. This configuration disappears when the tendon is stretched—here the collagen fibres straighten. When the stretching forces are removed, the tendon resumes its normal wavy appearance. If an acute stress causes an elongation of 8% or more, the tendon is likely to rupture [1].

Fibres along the tendon are not only parallel. Jozsa et al. [12] demonstrated that there are five types of fibre crossings—parallel running fibres, simply crossing fibres, crossing of two fibres with one straight running fibre, a plait formation with three fibres and an up-tying of parallel running fibres with one fibre. The ratio of longitudinal to transverse running fibres ranges between 10:1 and 26:1 [13]. Within one collagen fibre, the fibrils are oriented longitudinally and transversely.

The longitudinal fibres not only run parallel but also cross each other to form spirals [13].

The complex microstructure of the tendons correlates with their function to transmit the force created by the muscle to the bone and to make joint movement possible. During phases of various movements, the tendons are exposed to a number of forces—longitudinal, transversal and rotational as well as withstanding an array of pressures. Therefore, the internal structure of the tendon described serves as a buffer against forces of various directions, thus preventing damage and disconnection of the fibres [13].

3.3 Epitenon and endotenon

An entire flexor tendon is surrounded by a fine connective tissue sheath called the epitenon. Histologically, the epitenon consists of relatively dense network of collagen with strands of 8–10 nm in thickness [1]. It contains longitudinal, oblique and transverse fibrils. The outer surface of the epitenon is contiguous with the flexor sheath and inner surface with the endotenon. The endotenon resides inside the tendon; it invests each tendon fibre and also binds individual fibres as well as larger fibre bundles. In contrast to the epitenon, the endotenon consists of a thin reticular network of connective tissue inside the tendon with a crisscross pattern of collagen fibrils [1, 11]. The functions of the endotenon are to [10, 11]:

- Bind tendinous collagen fibres.
- Allow fibre groups to glide on each other.
- Carry blood vessels, nerves and lymphatics to the deeper portion of the tendon.

3.4 Tendon cells

Tendon cells are either tenoblasts or tenocytes which comprise 90–95% of the cells of the tendon [1]. The other 5–10% are chondrocytes (at the pressure and insertion sites), synovial cells of the tendon sheath (on the tendon surface) and vascular cells (capillary endothelial cells and smooth muscle cells of the arterioles). In pathological conditions, other cells can be observed in the tendon tissue such as inflammatory cells, macrophages and myofibroblasts [13].

Tenoblasts and tenocytes represent differing maturations of the tendon cell. Newborn tendons are called tenoblasts and have different shapes and sizes. In young individuals, the tenoblasts begin to resemble each other being spindle shaped. In adults, the cells are called tenocytes and are very elongated [13]. Tenoblasts and tenocytes are metabolically active cells and synthesise collagen and other matrix components [13, 14]. The metabolic pathways utilised for energy production change from aerobic to anaerobic with increasing age [10, 13]. The low metabolic rate of the tendon tissue, in addition to well-developed anaerobic energy production, is essential for the function of the tendon to carry loads and remain in tension for periods of time without the risk of ischaemia or necrosis [1]. A likely drawback of this low metabolic rate is the slow rate of recovery and healing after injury [15].

4. Flexor tendon healing: cellular concepts

In the 1960s, flexor tendon healing was thought to rely on the invasion of peripheral cells and blood vessels which lead to the formation of restrictive adhesions [16, 17]. This concept was contested over the next two decades when bodies of biologic and molecular evidence confirmed that tenocytes actively participate in tissue repair and that tendons are capable of healing from injury [18]. Tendon healing undergoes overlapping inflammation, proliferation and remodelling [19] via two mechanisms—extrinsic and intrinsic [8]. The proliferation of tenocytes and production of their extracellular matrix are the hallmark of the intrinsic process [20, 21]. Extrinsic healing on the other hand involves the invasion of fibroblasts and inflammatory cells into the site of injury from the surrounding synovium, paratenon and tendon sheath [8, 22].

4.1 Intrinsic healing

Intrinsic healing involves only the tenocytes (fibroblasts) within the tendon itself and depends on the migration and proliferation of cells from the epitenon and endotenon [8, 22]. Epitenon tenocytes produce collagen earlier than those of the endotenon. Tenocytes of the endotenon produce large and more mature collagen than epitenon cells. In any event, both endotenon and epitenon tenocytes establish an extracellular matrix and internal neovascular network. Intrinsic healing results in improved biomechanics within the sheath, including tendon gliding. Movement of the tendon within the sheath improves synovial circulation and therefore the delivery of nutrients.

4.2 Extrinsic healing

Extrinsic healing involves the invasion of fibroblasts and inflammatory cells into the site of injury from the surrounding synovium, paratenon and tendon sheath [8, 22]. This produces scarring and peritendinous adhesions which may impair tendon movement, gliding and nutrition (refer to Section 4.5). It is thought that extrinsic healing predominates in the earlier stages of tendon healing. Extrinsic healing also predominates when tendons are immobilised after injury or repair. The extrinsic mechanism is activated earlier and is responsible for initial adhesions, the highly cellular collagen matrix and the high-water content of the injury site [8, 22]. The intrinsic mechanism then causes tenocytes from within the tendon to invade the defect and produce collagen which reorganises and aligns longitudinally to maintain fibrillar continuity and produce a healed tendon [23].

4.3 Early healing stage

After tendon injury, two intricately related and balanced processes take place—tenocyte apoptosis and tenocyte proliferation [18, 24]. Wu et al. [25] specifically examined apoptosis and proliferation of a repaired digital flexor tendon in a chicken model. In uninjured tendons, only $3 \pm 2\%$ of the tenocytes showed signs of apoptosis, and $1 \pm 1\%$ showed signs of active proliferation. The percentage of apoptotic cells went up to more than 40% at days 3–7 after tendon injury; on day 3, the number of inflammatory cells in the wound site also peaked. The number of mainly inflammatory cells as well as tenocytes peaked during the very early days in the healing process (at days 3 and 7) in the chicken model. In addition, the number of proliferating cell nuclear antigen cells (PCNA) and Bcl-2 (an antiapoptotic protein) —markers of proliferation—did not significantly increase until day 7 and peaked during days 7–21. Thus, it was established that tenocyte apoptosis is accelerated within several days after injury, followed by increase in proliferation of tenocytes in 2–4 weeks with activation of molecular events to inhibit apoptosis.

4.4 Middle and late healing stages

Wu et al. then further quantified cell apoptosis and proliferation during the middle and late stages of healing [26]. The percentage of apoptotic tenocytes was generally higher on the surface of the tendon than that in the core, indicating a greater need for cellular clearance and surface remodelling in the surface region in the middle-to-late periods. Their findings also indicated that active tendon remodelling persists through the very late tendon healing period, especially on the surface because:

- The total cell population did not start to decline until after day 56 (2 months). The percentage of apoptotic tenocytes ranged from 30 to 40% in the total cell population.

- Cell apoptosis persisted at a relatively high level on the tendon surface at 3 months.

- Cell apoptosis in the core region declined after 2 months.

In sharp contrast to apoptosis, proliferation of tenocytes in the middle and late healing stages [26]:

- Declined drastically after week 4 (less than 5% of the PCNA-positive cells were found in the tendon).

- PCNA-positive cells were at normal levels at weeks 8–12.

The above two points indicated that tenocyte apoptosis is the dominant event in the middle and late tendon healing period.

In areas distant from the junction site, apoptosis is more prominent in the tendon surface than in the tendon core—this is thought to be associated with the clearance of excess cells, which serves to promote formation of smooth gliding surfaces by remodelling adhesions [26].

4.5 Adhesions

The primitive processes by which tissues repair after injury are indiscriminate to tissue types and lead to fibrotic scarring [27]. Flexor tendon tissue is not exempted from this. Injury to flexor tendons through trauma or surgery can result in problematic tendon adhesion formation. Adhesions affect the normal tendon gliding that occurs within a narrow flexor tendon sheath. Fibreoptic studies of surgical patients have demonstrated that the tendons, sheath, soft tissues and skin glide across each other in vascularised interconnecting tissue planes during finger flexion, with scarring of these planes affecting the fingers' ability to flex [28]. When two dynamic gliding planes are affected by injury, such as the tendon and sheath, the result is adhesions [28]. A landmark 1960 study by Lindsay and Thomson [29] had shown that immobilisation was key to adhesion formation after systematic wounding of the tendon, sheath, skin, soft tissue and vinculum complex. Further studies have shown that damage to the skin, sheath, soft tissues and vinculum alone is insufficient to form adhesions [28]. Additionally, keeping the damaged tendon and damaged soft tissue in relatively close approximation appears to be required for adhesions to form [29]. For these reasons, early active mobilisation is encouraged following tendon surgical repair [30].

Using a murine model, Wong et al. [28] demonstrated that the scarring between two damaged surfaces—i.e. the extrinsic healing process—was responsible for adhesions due to the increased inflammatory activity in the tissue surrounding the flexor tendons. They found that [28]:

- Adhesion formation was propagated by immobilisation of the digit.
- Inflammatory cells predominated in the surrounding tissues in the early phases of healing but appeared in the tendon proper during the remodelling phase.
- Proliferative activity occurred in both the surrounding tissues and tendon but was greater in the surrounding tissues.
- Collagen synthesis in the tendon and subcutaneous tissue is *temporally* different.
- Pericyte and myofibroblast activity predominates in the subcutaneous tissue and not tendon.
- In adhesion forming tendon wounds after 21 days, there were two distinct cell phenotypes observed. The first was a large cell with multiple cytoplasmic protrusions, which were seen to enclose large and small diameter fibrils or even multiple fibrils. The second phenotype was similar to those seen in the developing tendon, with small cytoplasmic protrusions and small fibrils being deposited by "fibripositors" (small fibrils in embryonic tendon fibroblasts which are enclosed in cytoplasmic processes). The number of fibripositors was greater than those seen in development.

What is clear is that the interactions between the damaged tissues and the processes that lead to adhesions are complex. The varying multicellular temporal and spatial expression involved in flexor tendon healing is far more intricate than that proposed by the intrinsic and extrinsic concepts of healing alone.

4.6 Microdynamics of adhesions at different stages of tendon healing

The mechanical characteristics of adhesion tissues determine tendon gliding, but this relationship is difficult to ascertain. Wu et al. [18] performed an in vivo study to determine the microdynamic features of adhesions in the middle and late healing periods (postoperative weeks 4–8). They found that the ability of adhesion tissues to resist tension decreased over time, whereas their flexibility increased; they hypothesised that this phenomenon determined the sliding amplitude of the tendon.

It was also found that, in a chicken toe flexor tendon that was surgically repaired and immobilised for 3 weeks, the percentage of apoptotic cell increased from the tendon core, to the tendon surface, to the adhesion-tendon interface and to the adhesion core [26]. Furthermore, tendons with more severe adhesions, i.e. those with less excursion, see greater apoptosis in their adhesions and adhesion-tendon interfaces.

In summary, it appears that the microdynamics of adhesions and tenocyte apoptosis are associated—*as apoptosis of the cells in the adhesions continues, the adhesions are more easily broken up after the adhesions are loaded* [18]. This would help, in part, explain why early active mobilisation is beneficial after tendon repair. Wu et al. [18] hypothesise that the external force applied to move the tendon during digital motion transfers to shear force over the adhesions and adhesion-tendon gliding interface. This continuously stimulates cellular apoptosis, in turn reducing the density and strength of the adhesion fibres, resulting in an increasingly greater elasticity and breakup of the adhesion fibres in the late healing stage.

4.7 Research trends

Recent research has examined methods to augment intrinsic biological healing of the flexor tendon whilst minimising adhesions.

- **Transforming growth factor β (TGF-β)**: Small amounts of TGF-β are found in the uninjured tendon [31]. The isoform TGF-β1 increases significantly after tendon injury [32, 33]. TGF-β1 has the highest association with adhesion formation and is therefore a major treatment target [31]. A neutralising antibody to TGF-β was shown to control scarring in rat dermal wounds [34, 35]. Furthermore, the same antibodies increased the total range of motion after flexor tendon repair in a rabbit model [36]. Additionally, the TGF-β1 receptor inhibitor SD208 can prevent progression and can improve tendon mechanical strength and decrease rupture rates [31]. However, suppression of TGF-β has been shown to decrease strength of tendon repair [37, 38]. This seems to be supported by gene therapy studies. Decreased TGF-β was examined by deleting the TGF-β inducible early gene (*Tieg1*)—this resulted in decreased collagen I deposition in an in vitro model of tendon healing [39].

- **Vascular endothelial growth factor (VEGF)**: It is known that tenocytes secrete VEGF and are present in synovial fibroblasts [8]. The VEGF family consists of several isoforms (VEGF-A, VEGF-B, VEGF-C, VEGF-D and VEGF-E and placenta growth factor), and these isoforms exert their effects through 3 tyrosine kinase receptors [31]. VEGF has been implicated in wound healing through epithelialization, collagen deposition and angiogenesis [40]. During flexor tendon repair, Boyer et al. showed that VEGF mRNA is increased during flexor tendon repair. It is postulated that the increased VEGF expression is associated with neovascularization [31]. VEGF genes delivered by adeno-associated virus (AAV) vectors in a chicken model demonstrated that healing strength was improved without increased adhesion formation [41].

- **Basic fibroblast growth factor (bFGF)**: bFGF, found within the tendon and tendon sheath, has been shown to influence wound healing due to its role in fibroblast chemotaxis, proliferation and angiogenesis [42]. However, its role in tendon healing remains unclear. Delivery of bFGF to injured tendons via adenoviral vector demonstrated improved tendon healing and increased strength with reduced adhesions [42].

- **Tissue engineering**: In their study, using a devitalised acellular allograft tendon containing recombinant AAV expressing growth and differentiation factor-5, Basile et al. [43] were able to repopulate the graft, decrease scar tissue and enhance the gliding property relative to the control graft. Tissue-engineered synovial membranes [44] have also been shown to decrease peritendinous adhesions.

5. Flexor tendon repair principles

This section will focus on repair principles. The reader is encouraged to seek alternative resources regarding specific repair techniques.

All suture tendon repair methods have been shown to significantly increase the gliding resistance compared to the intact tendon [45]. Gliding resistance is affected by [18, 46, 47]:

- The number of exposed suture loops and knots outside on the tendon surface
- The suture calibre
- The suture material
- Tendon bulkiness (from both oedema and surgical repair)
- Smoothness of tendon gliding surface
- The presence of intact annular pulleys
- Oedema
- Adhesions
- Joint stiffness
- Repaired flexor digitorum superficialis

Therefore, the ideal method of flexor tendon repair should allow a healing response precisely at the tendon ends but not between the tendon and its surroundings, create a repair site with minimal bulk and low friction and place enough force across the repair to promote motion and remodelling [22].

Strickland described the characteristics of an ideal tendon repair [48], and these were supported by further studies [22]. These are:

- Core sutures easily placed in tendon
- Secure knots
- Smooth junctions
- Minimal gapping
- Minimal interference with tendon vascularity
- Sufficient strength to permit application of early motion stress during the healing process
- Motion at the repair site to increase the amount of collagen deposited at the site of injury
- Equal tension across all suture strands

5.1 Repair strength

Initially, the strength of tendon repair depends solely on the repair technique [45]. It is postulated that postoperative tenomalacia may develop at the suture tendon junction, therefore decreasing initial repair strength [49]. The initial strength of the repair depends on the material properties and knot security of the sutures as well as on the holding capacity of the suture grips of the tendon [45]. Immobilisation significantly decreases the strength of repair within the first 3 weeks of healing [50], whereas early passive and early active motion have been shown to prevent the initial weakening, leading to progressively increased repair strength, starting from the time of repair [50–52]. The initial strength of the repair depends on the material of the suture itself, knot security of the suture and the holding capacity of the suture grips on the tendon [45]. Therefore, the biomechanical properties of the suture can be improved by:

- Increasing the number of strands crossing the repair site [53]
- Increasing the suture calibre [54]
- The number, size and configuration of the grips [45, 53]

5.2 Suture terminology

The flexor tendon repair is a composite of the *core* and *peripheral* sutures [55]. The core suture is the suture placed within the substance of the tendon proper and consists of at least two of three components—longitudinal, transverse and link. All core suture techniques have a longitudinal and link component.

- The link component is that part of the suture at the junction between longitudinal and the transverse components or between two longitudinal components. The link component lies outside the tendon.
- The longitudinal and transverse components are usually placed within the tendon substance, i.e. they are intratendinous.
- The transverse and/or link components convert the longitudinal pull of the suture to a transverse compressive force and prevent the longitudinal component from pulling out.
- The longitudinal component in turn allows placement of the transverse and/or link components away from the divided end of the tendon.

Pennington [56] first described the relationship of the transverse and longitudinal components when he outlined his locking-loop technique. *Locking* suture configurations tighten around bundles of tendon fibres with tension [56]; it can only do this when the transverse component crosses just superficial to the longitudinal part of the suture. The result is a loop of suture locking around a small bundle of tendon fibres so that when more tension is applied to the repair site, the tighter the grip of the suture loop on these fibre bundles [56]. *Grasping* loops on the other hand have the transverse component passing deep to the longitudinal constituent so that the suture does not pass around or lock a bundle of tendon fibres [57]. Locking loops improve the ultimate force and gap resistance compared to grasping loops in flexor tendon repair [45]. Several studies have demonstrated that locking loops improve the ultimate force and gap resistance compared to grasping loops in flexor tendon repair [45, 58]. However, the biomechanical advantage of the locking loops is obtained only with 3–0 or larger suture [45]. This is because with 4–0 suture, the material strength is inferior to the holding capacity of the suture grips of the tendon leading to failure by suture rupture before the true biomechanical properties of the locking loops are obtained [45]. Additionally, the size of the locking loop influences the biomechanical properties of the repair technique [59–61]. In the modified Pennington technique, increasing the cross-sectional area of each loop from 5 to 15% improved the ultimate force, whilst further increase did not improve strength, and the tendency for gap formation increased [60]. In the four-strand cruciate repair, the locking loops of 25% reached the highest gap force, ultimate force and stiffness [59].

Variations in the construction of the link component—arc, loop or knot—result in a *sliding* or an *anchored* suture on each half of the divided tendon [55].

- A sliding suture allows the suture to slide within the tendon substance when tension is applied to one of the longitudinal components. An arc link component results in sliding suture. When sliding sutures are used, tension is equally distributed among the different longitudinal strands.

- An anchored suture does not allow the suture to move independent of the tendon. A knot link component results in an anchored suture. When anchored sutures are used, the longitudinal strands are fixed. However, any slack in the suture will result in uneven distribution of tension and gapping at the tendon ends.

5.3 Suture principles

The length of the core suture purchase in the tendon logically determines how much of the segment of the tendon is incorporated into the repair. *The optimal range of core suture purchase has been determined as 1.0 cm with increased gap force, ultimate force and stiffness* [62, 63]. The purchase of 0.4 cm results in very weak repairs, whilst any increase over 1 cm does not improve the biomechanical properties [63].

Increasing the suture calibre has been shown to increase the ultimate force in static testing and fatigue strength in dynamic testing; however, it has not been shown to improve the yield force or gap resistance of the repairs [45]. The strength of the 4–0 suture has been reported to be less than the holding capacity of several locking and grasping repair techniques with failure occurring mostly by suture rupture [54, 64]. A 3–0 suture failure due to suture rupture and pullout has been reported [54, 64]. Therefore, the use of 3–0 suture is generally recommended to offer safety over the 4–0 suture by increasing the material strength [45, 54, 64].

The *ideal suture material* for flexor tendon repair should be strong enough; prevent gapping; be easy to use and knot; be absorbable but maintain its tensile properties until tendon repair has achieved adequate strength; and have minimal tissue response [65]. Non-absorbable, synthetic sutures, (especially coated braided polyester), monofilament nylon and monofilament polypropylene are used in flexor tendon repair [45]. Coated braided polyester suture is the most common core suture material, though nylon is also used, especially in repairs performed with looped suture. Monofilament polypropylene is mainly used in the peripheral sutures. Coated braided polyester suture demonstrates significantly higher tensile strength and stiffness than monofilament nylon and polypropylene sutures and maintains its tensile properties in the body temperature, whilst the stiffness of both polypropylene and nylon suture has been shown to decrease significantly [66, 67]. A braided polyblend polyethylene suture (Fiberwire®) has been introduced for flexor tendon repair. It has significantly higher ultimate force and stiffness than coated braided polyester, monofilament nylon and polypropylene sutures and a similar ultimate force but higher stiffness than braided stainless steel [66]. Bioabsorbable suture materials are not widely used in flexor tendon repair due to the lack of sufficient tensile strength half-life and potential increased tissue reaction and adhesion formation [45].

The original *peripheral or epitendinous* suture was thought of a "tidying up" suture to improve tendon gliding within the flexor sheath [68]. It has now been shown that the peripheral suture improves the gap resistance and strength of repair [45, 58]. The simple running peripheral suture is the most investigated and used technique in flexor tendon repair because of its simplicity [45]. The strength and stiffness of the running peripheral suture can be increased by:

- Taking deeper suture grasps [69]
- Increasing suture purchase from 1 to 2 or 3 mm [70]
- Increasing the number of suture passes [71]

The location and number of *knots* influence the strength of the tendon repair [72]. Ex vivo studies show that decreasing the number of knots and placing them outside the repair on the tendon surface increase the strength of the repair compared to knots placed between the tendon ends [45]. However, in in vivo studies, the knots placed inside the repair sites were stronger than those outsiders after 6 weeks [72].

6. Postoperative rehabilitation following flexor tendon repair

An understanding of the postoperative rehabilitation regimen after flexor tendon repair is of equal importance to the repair itself. Noncompliance with rehabilitation may lead to poor outcomes including repair rupture, decreased range of motion and joint stiffness. Current postoperative protocols for patients with flexor tendon injuries are immobilisation, early passive mobilisation and early active mobilisation [73, 74].

6.1 Immobilisation

The benefits of early mobilisation on the repair strength, tenocyte healing and formation of adhesions are widely known [22, 75–79]. Immobilisation, however, has its role in certain situations, particularly in patients who are noncompliant with early mobilisation protocols, paediatric patients, patients with cognitive deficits and patients with concurrent injuries that may be worsened with early active mobilisation (fractures, nerves and vessels) [73, 74].

It is difficult to encourage early mobilisation in children under the age of 6 [80]. O'Connell et al. showed outcomes were equal among children who were immobilised and those who underwent early mobilisation for 4 weeks [81]. However, immobilisation for more than 4 weeks resulted in functional deterioration of the repaired tendon [81].

The protocol of Cifaldi, Collins and Schwarze may be used for the noncompliant adult [73, 82]. This protocol involves 3–4 weeks of immobilisation in a forearm-based dorsal splint or cast (20° wrist flexion, MP joints in 50° flexion and the IP joints in full extension) followed by a weaning programme (it may also be used in children) [74, 82]. "Weaning" refers to modifying the splint in such a way that the wrist is in neutral and then instructing the patient to remove the splint every hour to passively flex and extend the injured digit for 10 repetitions. Splint wear is then discontinued at 6 weeks. From here, differential FDS and FDP gliding exercises are performed every hour for 10 repetitions [74]:

- To isolate FDP gliding, both the MP and PIP joints are held in extension, and the patient flexes the distal interphalangeal (DIP) joint. This prevents FDS glide.

- The FDS tendon glide exercise is achieved by isolating all fingers in extension, whilst the patient actively flexes the PIP joint of the affected finger. Holding the fingers in extension ensures that the common muscle belly of the FDP is held to its full length, preventing it from assisting in flexion.

- At postoperative week 8, sustained grip activities are added to the programme with resistance increasing over the next 4 weeks. Heavy resistive exercises are avoided before 12 weeks due to the risk of tendon rupture.

6.2 Early passive mobilisation

The inhibition of adhesion formation, promotion of intrinsic healing and production of a stronger repair can be encouraged with early passive mobilisation [77–79, 82–84]. The best known early passive mobilisation protocols are the Duran and Houser and Kleinert regimens [73, 74].

In the Duran and Houser protocol:

- A postoperative dorsal blocking splint holds the MP joints at 50° of flexion and the wrist at 20° of flexion. The following regimen is followed twice daily to ensure that 3–5 mm of tendon excursion occurs to prevent firm tendon adhesions [73].

- The patient uses the opposite hand to bring the PIP and the DIP joints from full flexion to full extension. This is done for eight repetitions for each joint.

- Then, the patient performs eight repetitions of composite MP, PIP and DIP flexion. The protocol continues through the fourth postoperative week.

- At 5 weeks, the patients begin active extension exercises with the use of a wristband. A rubber band is attached from the tip of the finger to the wristband, providing passive flexion and active extension. During this time, the patient also performs blocking and FDS gliding exercises.

- The late stage begins 8 weeks postoperatively. Progressive strength building is encouraged.

In the Kleinert protocol:

- A dorsal plaster splint is applied immediately at surgery. This splint blocks the wrist and MP joint in flexion. The wrist is placed at approximately 45° of flexion, the MP joints rest at approximately 20° of flexion and the IP joints are in neutral.

- One week following surgery, the plaster is replaced with a thermoplastic splint that maintains the same flexion angles as above.

- The new splint allows for passive flexion of the digits and active extension of the digits against dynamic traction using rubber bands to facilitate the traction mechanism. These bands are placed on the volar aspect of the splint and directed towards the distal nail plate from just proximal to the wrist.

- Early passive ROM exercises are started within the dorsal splint.

- At 1 month, the splint is removed, and active flexion and extension exercises begin. However, the dorsal splint must be worn when these exercises are not being performed.

- At 6 weeks, the dorsal splint is discontinued, and blocking exercises commence.

- Two months following the repair, resistive exercises are incorporated into the regimen.

- Resumption of normal activities occurs approximately 3 months following the surgical repair.

A major issue with the Kleinert protocol is the development of flexion contractures of the PIP joint [85]. These can be treated with continued intermittent splinting of the IP joints in neutral [86]. In recent years, rubber band traction has been almost completely abandoned, largely because of the problems arising from the flexed resting position of the PIP joint [87].

Continuous passive motion (CPM) uses devices that allow for joints to move through a predetermined arc of motion [73]. The goal is to increase the duration and repetition of exercises. A randomised control comparing traditional early passive motion to CPM exercises [88] showed that, at 6 months, the CPM group had significantly greater range of motion. However, further research in evaluating the CPM following flexor tendon repair is lacking.

6.3 Early active mobilisation

An early active mobilisation (EAM) protocol refers to active contraction of the repaired muscles [89, 90]. EAM has been shown to promote the formation of large diameter fibrils, and it demonstrates the greatest cellular response to injury [83]. There are many different EAM regimens in the literature [91, 92]. Gratton [93] combined the Belfast and Sheffield practices [89] to form a widely used regimen:

- A thermoplastic dorsal blocking splint is applied at postoperative day 2–5 with the wrist positioned in 20° of flexion, the MCP joints in 80° of flexion and the IP joints in full extension. Active ROM exercises are delayed until day 5 if there is significant oedema which should be treated with compression and elevation.

- In the absence of significant oedema, exercises begin with passive flexion of the digits and active extension to the constraints of the dorsal splint.

- At the completion of the above exercises, active flexion exercises begin. Here, a finger of the opposite hand is placed in the palm of the affected hand, and the patient flexes the affected fingers against the contralateral fingers aiming to progress one finger width per week.

- By the end of week 1, the patient is expected to have full passive flexion, full active extension and PIP active flexion to 30°.

- Discontinuation of the splint occurs between weeks 4 and 6—week 4 for patients with poor tendon gliding and 6 for those who have excellent ROM (defined as full active fist at week 2). At this time, exercises consist of passive ROM and active ROM.

- From week 6, blocking exercises of the individual joint is commenced. If flexion contractures are evident, these will need to be corrected with a splint.

- Progressive strengthening exercises begin 3 weeks after the dorsal block splint is discontinued. Resistance should increase so as to allow the patient to have full hand function by week 12.

It should be noted that EAM protocols should be *individualised* to the patient [73, 94] because advancement to the next phase of a protocol may be hindered or augmented based on the level of oedema, passive versus active flexion lags and adhesion formation [94].

Irrespective of whether or not a passive or active protocol is used, it has been shown that initiating a mobilisation therapy by postoperative day 5 decreases the rate of secondary procedures and decreases the costs of treatment [95].

7. Conclusion

The fine, tailored movements of the flexor tendon are essential to hand function. It is clear that the consequences of extrinsic healing of flexor tendons must be overcome to achieve optimal outcomes in patients who have injured their flexor tendons. Until the intrinsic healing process can be biologically augmented, surgical repair and rehabilitation of the injured flexor tendon will remain the mainstays of treatment. It is therefore essential that the surgeon bear in mind the basic tenets of tendon healing and the foundational principles of surgical repair.

Acknowledgements

The author wishes to acknowledge the immense contribution to flexor tendon pathology by the researchers, scientists and clinicians cited in this chapter.

Conflict of interest

None to declare.

Author details

Justin Yousef
Department of Plastic and Reconstructive Surgery, The Canberra Hospital, Canberra, Australia

*Address all correspondence to: justin.yousef@gmail.com

References

[1] Kannus P. Structure of the tendon connective tissue. Scandinavian Journal of Medicine and Science in Sports. 2000;**10**(6):312-320

[2] Schöffl V, Heid A, Küpper T. Tendon injuries of the hand. World Journal of Orthopedics. 2012;**3**(6):62

[3] Zafonte B, Rendulic D, Szabo RM. Flexor pulley system: Anatomy, injury, and management. The Journal of Hand Surgery. 2014;**39**(12):2525-2532

[4] Mitsionis G, Bastidas JA, Grewal R, Pfaeffle HJ, Fischer KJ, Tomaino MM. Feasibility of partial A2 and A4 pulley excision: Effect on finger flexor tendon biomechanics. The Journal of Hand Surgery. 1999;**24**(2):310

[5] Tomaino M, Mitsionis G, Basitidas J, Grewal R, Pfaeffle J. The effect of partial excision of the A2 and A4 pulleys on the biomechanics of finger flexion. The Journal of Hand Surgery. 1998;**23**(1): 50-52

[6] Sherman PJ, Lane LB. The palmar aponeurosis pulley as a cause of trigger finger. A report of two cases. JBJS. 1996; **78**(11)

[7] Schubert MF, Shah VS, Craig CL, Zeller JL. Varied anatomy of the thumb pulley system: Implications for successful trigger thumb release. The Journal of Hand Surgery. 2012;**37**(11): 2278-2285

[8] Myer C, Fowler JR. Flexor tendon repair: Healing, biomechanics, and suture configurations. The Orthopedic Clinics of North America. 2016;**47**(1): 219-226

[9] Ochiai N, Matsui T, Miyaji N, Merklin RJ, Hunter JM. Vascular anatomy of flexor tendons. I. Vincular system and blood supply of the profundus tendon in the digital sheath. The Journal of Hand Surgery. 1979;**4**(4): 321-330

[10] Hess GP, Cappiello WL, Poole RM, Hunter SC. Prevention and treatment of overuse tendon injuries. Sports Medicine (Auckland, NZ); 1989;**8**(6): 371-384

[11] Rowe RWD. The structure of rat tail tendon fascicles. Connective Tissue Research. 1985;**14**(1):21-30

[12] Józsa L, Réffy A, Bálint JB. Polarization and electron microscopic studies on the collagen of intact and ruptured human tendons. Acta Histochemica. 1984;**74**(2):209, ins1,15-14,ins1,15

[13] Jozsa L, Kannus P, Balint JB, Reffy A. Three-dimensional infrastructure of human tendons. Cells, Tissues, Organs. 1991;**142**(4):306-312

[14] O'Brien M. Structure and metabolism of tendons. Scandinavian Journal of Medicine and Science in Sports. 1997;7(2):55-61

[15] Vailas AC, Tipton CM, Laughlin HL, Tcheng TK, Matthes RD. Physical activity and hypophysectomy on the aerobic capacity of ligaments and tendons. Journal of Applied Physiology. 1978;**44**(4):542-546

[16] Potenza AD. Critical evaluation of flexor-tendon healing and adhesion formation within artificial digital sheaths: An experimental study. JBJS. 1963;**45**(6):1217-1233

[17] Potenza AD. Tendon healing within the flexor digital sheath in the dog: An experimental study. JBJS. 1962;**44**(1): 49-64

[18] Wu YF, Tang JB. Tendon healing, edema, and resistance to flexor tendon

gliding: Clinical implications. Hand Clinics. 2013;**29**(2):167-178

[19] Gelberman RH, Vandeberg JS, Manske PR, Akeson WH. The early stages of flexor tendon healing: A morphologic study of the first fourteen days. The Journal of Hand Surgery. 1985;**10**(6 Pt 1):776-784

[20] Ingraham JM, Weber RA, Childs EW. Intrinsic tendon healing requires the recycling of tendon collagen fibril segments. The Journal of Hand Surgery (European Volume). 2011;**36**(2):154-155

[21] Jones M, Mudera V, Brown R, Cambrey A, Grobbelaar A, Angus McGrouther D. The early surface cell response to flexor tendon injury. The Journal of Hand Surgery (Am). 2003;**28**: 221-230

[22] Beredjiklian PK. Biologic aspects of flexor tendon laceration and repair. The Journal of Bone and Joint Surgery. American Volume. 2003;**85-A**(3): 539-550

[23] James R, Kesturu G, Balian G, Chhabra AB. Tendon: Biology, biomechanics, repair, growth factors, and evolving treatment options. The Journal of Hand Surgery. 2008;**33**(1): 102-112

[24] Lui PPY, Cheuk YC, Hung LK, Fu SC, Chan KM. Increased apoptosis at the late stage of tendon healing. Wound Repair and Regeneration. 2007;**15**(5): 702-707

[25] Wu YF, Chen CH, Cao Y, Avanessian B, Wang XT, Tang JB. Molecular events of cellular apoptosis and proliferation in the early tendon healing period. The Journal of Hand Surgery. 2010;**35**(1):2-10

[26] Wu YF, Zhou YL, Mao WF, Avanessian B, Liu PY, Tang JB. Cellular apoptosis and proliferation in the middle and late intrasynovial tendon healing periods. The Journal of Hand Surgery. 2011;**37**(2):209-216

[27] Ferguson MWJ, O'Kane S. Scar-free healing: From embryonic mechanisms to adult therapeutic intervention. Philosophical Transactions of the Royal Society of London. Series B, Biological Sciences. 2004;**359**(1445): 839-850

[28] Wong JKF, Lui YH, Kapacee Z, Kadler KE, Ferguson MWJ, McGrouther DA. The cellular biology of flexor tendon adhesion formation: An old problem in a new paradigm. The American Journal of Pathology. 2009; **175**(5):1938-1951

[29] Lindsay WK, Thomson HG. Digital flexor tendons: An experimental study. Part I. The significance of each component of the flexor mechanismhin tendon healing. British Journal of Plastic Surgery. 1960;**12**:289

[30] Small JO, Brennen MD, Colville J. Early active mobilisation following flexor tendon repair in zone 2. The Journal of Hand Surgery: British & European Volume. 1989;**14**(4):383-391

[31] Legrand A, Kaufman Y, Long C, Fox PM. Molecular biology of flexor tendon healing in relation to reduction of tendon adhesions. The Journal of Hand Surgery. 2017;**42**(9):722-726

[32] Chang J. Studies in flexor tendon reconstruction: Biomolecular modulation of tendon repair and tissue engineering. The Journal of Hand Surgery. 2012;**37**(3):552-561

[33] Chang J, Most D, Stelnicki E, Siebert JW, Longaker MT, Hui K, et al. Gene expression of transforming growth factor beta-1 in rabbit zone II flexor tendon wound healing: Evidence for dual mechanisms of repair. Plastic and Reconstructive Surgery. 1997; **100**(4):937

[34] Shah M, Foreman DM, Ferguson MW. Neutralisation of TGF-beta 1 and TGF-beta 2 or exogenous addition of TGF-beta 3 to cutaneous rat wounds reduces scarring. Journal of Cell Science. 1995;**108**(Pt 3):985

[35] Shah M, Foreman DM, Ferguson MWJ. Control of scarring in adult wounds by neutralising antibody to transforming growth factor β. The Lancet. 1992;**339**(8787):213-214

[36] Chang J, Thunder R, Most D, Longaker MT, Lineaweaver WC. Studies in flexor tendon wound healing: Neutralizing antibody to TGF-beta1 increases postoperative range of motion. Plastic and Reconstructive Surgery. 2000;**105**(1):148

[37] Loiselle AE, Kelly M, Hammert WC. Biological augmentation of flexor tendon repair: A challenging cellular landscape. The Journal of Hand Surgery. 2016;**41**(1):144-149; quiz 9

[38] Zhou Y, Zhang L, Zhao W, Wu Y, Zhu C, Yang Y. Nanoparticle-mediated delivery of TGF-β1 miRNA plasmid for preventing flexor tendon adhesion formation. Biomaterials. 2013;**34**(33): 8269-8278

[39] Tsubone T, Moran SL, Subramaniam M, Amadio PC, Spelsberg TC, An KN. Effect of TGF-β inducible early gene deficiency on flexor tendon healing. Journal of Orthopaedic Research. 2006;**24**(3):569-575

[40] Barrientos S, Stojadinovic O, Golinko MS, Brem H, Tomic-Canic M. Perspective article: Growth factors and cytokines in wound healing. Wound Repair and Regeneration. 2008;**16**(5): 585-601

[41] Mao WF, Wu YF, Yang QQ, Zhou YL, Wang XT, Liu PY, et al. Modulation of digital flexor tendon healing by vascular endothelial growth factor gene transfection in a chicken model. Gene Therapy. 2017;**24**(4):234-240

[42] Tang BJ, Cao TY, Zhu YB, Xin YK-Q, Wang YX, Liu YP. Adeno-associated virus-2-mediated bfgf gene transfer to digital flexor tendons significantly increases healing strength: An in vivo study. The Journal of Bone and Joint Surgery. 2008;**90**(5):1078-1089

[43] Basile P, Dadali T, Jacobson J, Hasslund S, Ulrich-Vinther M, Søballe K, et al. Freeze-dried tendon allografts as tissue-engineering scaffolds for Gdf5 gene delivery. Molecular Therapy. 2008; **16**(3):466-473

[44] Baymurat AC, Ozturk AM, Yetkin H, Ergun MA, Helvacıoglu F, Ozkızılcık A, et al. Bio-engineered synovial membrane to prevent tendon adhesions in rabbit flexor tendon model: Bio-engineered synovial membrane. Journal of Biomedical Materials Research, Part A. 2015;**103**(1):84-90

[45] Viinikainen A, Göransson H, Ryhänen J. Primary flexor tendon repair techniques. Scandinavian Journal of Surgery. 2008;**97**(4):333-340

[46] Zhao C, Amadio PC, Zobitz ME, Momose T, Couvreur P, An K-N. Gliding resistance after repair of partially lacerated human flexor digitorum profundus tendon in vitro. Clinical Biomechanics. 2001;**16**(8): 696-701

[47] Momose T, Amadio PC, Zhao C, Zobitz ME, An KN. The effect of knot location, suture material, and suture size on the gliding resistance of flexor tendons. Journal of Biomedical Materials Research. 2000;**53**(6):806-811

[48] Rudge WBJ, James M. Flexor tendon injuries in the hand: A UK survey of repair techniques and suture materials—Are we following the evidence? ISRN Plastic Surgery. 2014; **2014**:4

[49] McDowell CL, Marqueen TJ, Yager D, Owen J, Wayne JS. Characterization of the tensile properties and histologic/biochemical changes in normal chicken tendon at the site of suture insertion. The Journal of Hand Surgery. 2002;**27**(4): 605-614

[50] Hitchcock TF, Light TR, Bunch WH, Knight GW, Sartori MJ, Patwardhan AG, et al. The effect of immediate constrained digital motion on the strength of flexor tendon repairs in chickens. The Journal of Hand Surgery. 1987;**12**(4):590-595

[51] Wada A, Kubota H, Miyanishi K, Hatanaka H, Miura H, Iwamoto Y. Comparison of postoperative early active mobilization and immobilization in vivo utilising a four-strand flexor tendon repair. The Journal of Hand Surgery (Br). 2001;**26**:301-306

[52] Wada A, Kubota H, Akiyama T, Hatanaka H, Miura H, Iwamoto Y. Effect of absorbable polydioxanone flexor tendon repair and restricted active mobilization in a canine model. The Journal of Hand Surgery. 2001; **26**(3):398-406

[53] Savage R. In vitro studies of a new method of flexor tendon repair. The Journal of Hand Surgery: British & European Volume. 1985;**10**(2):135-141

[54] Taras JS, Raphael JS, Marczyk SC, Bauerle WB. Evaluation of suture caliber in flexor tendon repair. The Journal of Hand Surgery. 2001;**26**(6): 1100-1104

[55] Sebastin SJ, Ho A, Karjalainen T, Chung KC. History and evolution of the Kessler repair. The Journal of Hand Surgery. 2013;**38**(3):552-561

[56] Pennington GD. The locking loop tendon suture. Plastic and Reconstructive Surgery. 1979;**63**(5): 648-652

[57] Hotokezaka S, Manske PR. Differences between locking loops and grasping loops: Effects on 2-strand core suture. The Journal of Hand Surgery. 1997;**22**(6):995-1003

[58] Rawson S, Cartmell S, Wong J. Suture techniques for tendon repair; a comparative review. Muscles, Ligaments & Tendons Journal (MLTJ). 2013;**3**(3):220-228

[59] Dona E, Turner AWL, Gianoutsos MP, Walsh WR. Biomechanical properties of four circumferential flexor tendon suture techniques 11 No benefits in any form have been received or will be received by a commercial party related directly or indirectly to the subject of this article. The Journal of Hand Surgery. 2003; **28**(5):824-831

[60] Hatanaka H, Manske PR. Effect of the cross-sectional area of locking loops in flexor tendon repair. The Journal of Hand Surgery. 1999;**24**(4):751-760

[61] Xie RG, Xue HG, Gu JH, Tan J, Tang JB. Effects of locking area on strength of 2- and 4-strand locking tendon repairs. The Journal of Hand Surgery. 2005; **30**(3):455-460

[62] Cao Y, Zhu B, Xie RG, Tang JB. Influence of core suture purchase length on strength of four-strand tendon repairs. The Journal of Hand Surgery. 2006;**31**(1):107-112

[63] Tang JB, Zhang Y, Cao Y, Xie RG. Core suture purchase affects strength of tendon repairs. The Journal of Hand Surgery. 2005;**30**(6):1262-1266

[64] Barrie K, Tomak S, Cholewicki J, Merrell G, Wolfe S. Effect of suture locking and suture caliber on fatigue strength of flexor tendon repairs The Journal of Hand Surgery (Am). 2001;**26**: 340-346

[65] Trail IA, Powell ES, Noble J. An evaluation of suture materials used in

tendon surgery. The Journal of Hand Surgery: British & European Volume. 1989;**14**(4):422-427

[66] Lawrence TM, Davis TRC. A biomechanical analysis of suture materials and their influence on a four-strand flexor tendon repair. The Journal of Hand Surgery. 2005;**30**(4):836-841

[67] Vizesi F, Jones C, Lotz N, Gianoutsos M, Walsh WR. Stress relaxation and creep: Viscoelastic properties of common suture materials used for flexor tendon repair. The Journal of Hand Surgery. 2008;**33**(2): 241-246

[68] Lister GD, Kleinert HE, Kutz JE, Atasoy E. Primary flexor tendon repair followed by immediate controlled mobilization. The Journal of Hand Surgery (Am). 1977;**6**:441-451

[69] Diao E, Hariharan JS, Soejima O, Lotz JC. Effect of peripheral suture depth on strength of tendon repairs. The Journal of Hand Surgery. 1996;**21**(2): 234-239

[70] Merrell G, Wolfe S, J Kacena W, Gao Y, Cholewicki J, Kacena M. The effect of increased peripheral suture purchase on the strength of flexor tendon repairs The Journal of Hand Surgery (Am). 2003; **28**:464-468

[71] Kubota H, Aoki M, Pruitt DL, Manske PR. Mechanical properties of various circumferential tendon suture techniques. The Journal of Hand Surgery: British & European Volume. 1996;**21**(4):474-480

[72] Aoki M, Pruitt DL, Kubota H, Manske PR. Effect of suture knots on tensile strength of repaired canine flexor tendons. The Journal of Hand Surgery: British & European Volume. 1995;**20**(1): 72-75

[73] Baskies MA, Tuckman DV, Paksima N. Management of flexor tendon injuries following surgical repair. Bulletin of the NYU Hospital for Joint Diseases. 2008;**66**(1):35-40

[74] Yousef J, Anthony S. Flexor tendon injuries. In: Salgado AA, editor. Essentials of Hand Surgery. Rijeka, Croatia: IntechOpen; 2018

[75] Mason ML, Allen HS. The rate of healing of tendons: An experimental study of tensile strength. Annals of Surgery. 1941;**113**(3):424-459

[76] Birdsell DC, Tustanoff ER, Kindsay WK. Collagen production in regenerating tendon. Plastic and Reconstructive Surgery. 1966;**37**(6): 504-511

[77] Gelberman RH, Menon J, Gonsalves M, Akeson WH. The effects of mobilization on the vascularization of healing flexor tendons in dogs. Clinical Orthopaedics and Related Research. 1980;(153):283-289

[78] Woo SL, Gelberman RH, Cobb NG, Amiel D, Lothringer K, Akeson WH. The importance of controlled passive mobilization on flexor tendon healing. A biomechanical study. Acta Orthopaedica Scandinavica. 1981;**52**(6):615-622

[79] Woo SL, Gomez MA, Amiel D, Ritter MA, Gelberman RH, Akeson WH. The effects of exercise on the biomechanical and biochemical properties of swine digital flexor tendons. Journal of Biomechanical Engineering. 1981;**103**(1):51-56

[80] Kato H, Minami A, Suenaga N, Iwasaki N, Kimura T. Long-term results after primary repairs of zone 2 flexor tendon lacerations in children younger than age 6 years. Journal of Pediatric Orthopaedics. 2002;**22**(6):732-735

[81] O'Connell SJ, Moore MM, Strickland JW, Thomas Frazier G, Dell PC. Results of zone I and zone II flexor tendon repairs in children. The Journal of Hand Surgery. 1994;**19**(1):48-52

[82] Collins DC, Schwarze L. Early progressive resistance following immobilization of flexor tendon repairs. Journal of Hand Therapy. 1991;**4**(3): 111-116

[83] Kubota H, Manske PR, Aoki M, Pruitt DL, Larson BJ. Effect of motion and tension on injured flexor tendons in chickens. The Journal of Hand Surgery. 1996;**21**(3):456

[84] Moriya K, Yoshizu T, Maki Y, Tsubokawa N, Narisawa H, Endo N. Clinical outcomes of early active mobilization following flexor tendon repair using the six-strand technique: Short- and long-term evaluations. The Journal of Hand Surgery, European Volume. 2015;**40**(3):250-258

[85] May EJ, Silfverskiöld KL, Sollerman CJ. Controlled mobilization after flexor tendon repair in zone II: A prospective comparison of three methods. The Journal of Hand Surgery. 1992;**17**(5): 942-952

[86] Stegink Jansen CW, Minerbo G. A comparison between early dynamically controlled mobilization and immobilization after flexor tendon repair in zone 2 of the hand: Preliminary results. Journal of Hand Therapy. 1990; **3**(1):20-25

[87] Elliot D, Giesen T. Avoidance of unfavourable results following primary flexor tendon surgery. Indian Journal of Plastic Surgery: Official Publication of the Association of Plastic Surgeons of India. 2013;**46**(2):312-324

[88] Gelberman RH, Nunley nJA, Osterman AL, Breen TF, Dimick MP, Woo SL. Influences of the protected passive mobilization interval on flexor tendon healing. A prospective randomized clinical study. Clinical Orthopaedics and Related Research. 1991;(264):189-196

[89] Small JO, Brennen MD, Colville J. Early Active Mobilisation Following Flexor Tendon Repair in Zone 2. Scotland: Elsevier Ltd; 1989. pp. 383-391

[90] Allen BN, Frykman GK, Unsell RS, Wood VE. Ruptured flexor tendon tenorrhaphies in zone II: Repair and rehabilitation. The Journal of Hand Surgery. 1987;**12**(1):18-21

[91] Tang JB, Amadio PC, Boyer MI, Savage R, Zhao C, Sandow M, et al. Current practice of primary flexor tendon repair: A global view. Hand Clinics. 2013;**29**(2):179. U6 - ctx_ver= Z3988–2004&ctx_enc=info%3Aofi%2Fenc%3AUTF-8&rfr_id=info%3Asid%2Fsummonserialssolutionscom&rft_val_fmt=info%3Aofi%2Ffmt%3Akev%3Amtx%3Ajournal&rftgenre=article&rftatitle=Current+practice+of+primary+flexor+tendon+repair%3A+a+global+view&rftjtitle=Hand+clinics&rftau=Tang%2C+Jin+Bo&rftau=Amadio%2C+Peter+C&rftau=Boyer%2C+Martin+I&rftau=Savage%2C+Robert&rftdate=2013-05-01&rfteissn=1558-1969&rftvolume=29&rftissue=2&rftspage=179&rft_id=info%3Apmid%2F23660054&rftexternalDocID=23660054&paramdict=en-US U7

[92] Pettengill KM. The evolution of early mobilization of the repaired flexor tendon. Journal of Hand Therapy. 2005; **18**(2):157-168

[93] Gratton P. Early active mobilization after flexor tendon repairs. Journal of Hand Therapy. 1993;**6**(4):285-289

[94] Kannas S, Jeardeau TA, Bishop AT. Rehabilitation following zone II flexor tendon repairs. Techniques in Hand & Upper Extremity Surgery. 2015;**19**(1): 2-10

[95] Hsiao P-C, Yang S-Y, Ho C-H, Chou W, Lu S-R. The benefit of early rehabilitation following tendon repair of the hand: A population-based claims database analysis. Journal of Hand Therapy: Official Journal of the American Society of Hand Therapists. 2015;**28**(1):20-26

Chapter 8

Management of Flexor Tendon Injuries in Hand

Tawheed Ahmad, Summaira Jan and Saima Rashid

Abstract

Peculiar anatomy of human hand with flexing fingers and opposing thumb give human beings clear edge over other existing living beings. We use hands for carrying out most of our daily activities. But at the same time this makes our hands vulnerable for getting traumatized. Hand injuries which involve underlying tendons make digits dysfunctional, which in turn affects overall precise functioning of hand. In this chapter we will briefly discuss related surgical anatomy of flexor tendons and associated structures, features of flexor tendon injuries at different zonal levels, surgical methods involved and different post-operative protocols used for management of these flexor tendon injuries.

Keywords: flexor tendon injuries, flexor digitorum superficialis, hand injuries, flexor digitorum profundus, tenorrhaphy

1. Introduction

Philosophically, physiologically and anatomically, the interaction of the brain and the hand give unique identification to *Homo sapiens*. The progress of mankind has been credited for large extent to evolution of mobile and strong upper limb having independently opposing thumb with cognitive power of using it. The precision, balance and its specialization give human hand a central functional as well as communicative role. The aim of surgical treatment for the injured, diseased or dysfunctional hand is to retain its maximal useful length, stable motion and unimpaired mobility of sensate parts.

All functions of hand are executed with the help of digits, and tendons in turn execute the movement of digits.

Flexion of fingers is done by two tendons viz. flexor digitorum superficialis (FDS) and flexor digitorum profundus (FDP). Flexor digitorum superficial splits into radial and ulnar slips prior to insertion on middle phalanx. Flexor digitorum profundus passes between the two slips of flexor digitorum superficialis, through a space called "Camper's chiasm". Flexor digitorum profundus is inserted on volar aspect, near the base, of distal phalanx of the respective finger.

The Flexor digitorum superficialis and profundus tendons glide together in a fibro-osseous tunnel. This fibro-osseous tunnel is composed of five annular pulleys (A_1–A_5) and three cruciate pulleys (C_1–C_3). These pulleys prevent bowstringing and the increase mechanical effectiveness of pull across the joints.

Annular pulleys A_2 and A_4 pulleys are the most critical to finger function and are located in proximal part of proximal phalanx and middle part of middle phalanx, respectively.

Annular pulleys A_1, A_3 and A_5 pulleys are located at the metacarpo-phalangeal, proximal interphalangeal and distal interphalangeal joints, respectively. Three cruciate pulleys C_1, C_2 and C_3 are located between A_2–A_3, A_3–A_4 and A_4–A_5 pulleys, respectively.

Thumb has three pulleys A_1, oblique and A_2 pulley located over metacarpo-phalangeal joint, proximal phalanx and inter-phalangeal joint respectively. The oblique pulley is the most mechanically important pulley among them.

Flexor tendons of hand are divided into five zones:

Zone I: Extends from finger top to insertion of flexor digitorum superficialis.

Zone II: This extends from insertion of flexor digitorum superficialis up to distal palmar crease.

Zone III: Extends from distal palmar crease up to flexor retinaculum.

Zone IV: This zone lies under flexor retinaculum.

Zone V: Extends from proximal border of flexor retinaculum to musculo-tendinous junction of flexor muscles.

Flexor tendon within fibro-osseous canal (zone II) receives nutrition from two distinct sources i.e. vascular and synovial. Four digital arches are formed by the anastomosis of branches from the two digital arteries. These arches are located at the base and the neck of proximal and middle phalanx. A vinculum arises from each of these arches (V_1–V_4). Vinculae V_1 and V_2 supply the flexor digitorum superficialis, whereas V_3 and V_4 supply flexor digitorum profundus tendon.

The surface of the tendon that is not compressed during flexion is supplied by perfusion with arterial blood, the surface that is compressed i.e. the palmar surface, is supplied by diffusion of synovial fluid. Diffusion is a more significant nutritive pathway than perfusion. The exact proportions of the two have been estimated to be 2:1 in flexor digitorum superficialis, and 5:1 in the flexor digitorum profundus.

Microscopically, tendons are composed of collagen bundles (mainly type I) oriented in regular, spiraling pattern with very few tendon cells (tenocytes), synovial cells and fibroblasts. 'Endotenon' encloses tendon bundles. If the tendon is within a synovial sheath, the outer layer of tendon is called the 'Epitenon' and if the tendon is outside the sheath (extra-synovial), outer loose areolar adventitial layer is called the 'Paratenon', through which blood vessels run longitudinally.

Tendons heal by intrinsic and extrinsic mechanism. Extrinsic mechanism of healing is by fibroblast cells in surrounding tissues and is responsible for adhesion formation, while as intrinsic mechanism of healing is because of tenocytes present within the tendon. Tendon gliding exercises after tendon repair promotes intrinsic mechanism of healing and inhibits the extrinsic mechanism of healing, thus preventing post tendon repair adhesion formation.

Flexor tendon injuries are common but a difficult problem for the patient, hand surgeon and the therapist. The incidence of flexor tendon injuries in industrialized countries is estimated to be 1 in 7000. The impact to patients may include loss of function, stiffness, vocational impairment and associated social and economical hardships.

1.1 Pulley system

The synovial sheath is reinforced by a fibrous pulley system delimiting the digital canal. This system is composed of five annular pulleys and three cruciform pulleys [1–3]. The first, third and fifth annular pulleys arise respectively from the volar plate of the metacarpophalangeal (MCP) joint, proximal interphalangeal (PIP) joint and distal interphalangeal (DIP) joint. The remaining two annular pulleys second and the fourth one arise respectively from the proximal and middle phalanx; these two pulleys are thicker and broader than the rest. Between the second and third annular pulleys is located the first cruciform pulley; between the third and fourth annular pulleys is present the second cruciform pulley; and the third cruciform pulley is located just proximal to the DIP joint. The cruciform pulleys play a role in the production of synovial fluid. The palmar aponeurosis pulley located at distal part of the transverse fibers of the palmar aponeurosis, just close to the beginning of the membranous synovial sheath, should be considered part of the finger pulley system; on each side of the synovial sheath the vertical fibers anchor it to the deep transverse metacarpal ligament. The digital fibrous pulley system keeps the flexor tendons close to the bone thus allowing complete flexion of the finger. The importance of palmar aponeurosis pulley increases significantly in case of absence of any annular pulley. The palmar aponeurosis having breaking strength of 16.5 kg is superior to the A2 pulley with breaking strength of 14 kg [4]. If A2 and A4 pulleys are absent there is significant loss of finger flexion.

1.2 Healing process of tendons

The two processes are involved in the healing process of tendons: the extrinsic healing mechanism involving the surrounding tissues, and the intrinsic healing mechanism, that involves the tendon itself and its synovial sheath. Vascular and cellular ingrowths from the surrounding tissues enhance the extrinsic healing. The callus formed allows the cicatrization of the tendon but at the same time restricts its mobility, especially in zone II. To prevent formation of these adhesions, agents like steroids, anti-inflammatory drugs, hyaluronic acid and anti-histaminics have been proposed [5, 6]. However, for decreasing the risk of adhesion formation microsurgical techniques and new suture materials along with an atraumatic approach has been very effective. But many factors, such as associated lesions (skin loss, vascular, nerve injury or fracture) and the nature of the trauma (avulsion, crush injuries, blunt injury) play a significant role in increasing chance of adhesion formation. Gelberman and colleagues [7] reported the benefits of gliding function and protected passive mobilization on the tensile strength as compared to complete immobilization of repaired tendons.

Studies have shown that the tendon cells (tenocytes) themselves have a potential of healing. Lundborg et al. [8] showed that a flexor tendon that is isolated and kept in a synovial fluid environment, without any vascular supply, is able to survive and heal without any formation of adhesions. Therefore, its emphasized that during process of tendon repair, as much as possible synovial sheath should be preserved [9–11].

2. Surgical techniques for primary flexor tendon repair

2.1 General considerations

Usually flexor tendon repair is performed in an emergency setup. In cases such as dirty trauma or crush injuries, debridement should be done to convert contaminated wound into a cleaner wound. All the injuries (fracture, skin loss, neurovascular bundles) are repaired simultaneously along with flexor tendon repair. However, if the surgeon does not possess enough expertise to treat such lesions, it's advisable to delay the repair till next appropriate time [12]. All injured flexor tendons should be repaired using proper instruments and under magnification in an operating room thus allowing atraumatic repair of such tendons. Cleaning of wound before tenorrhaphy and in certain circumstances administration of intravenous antibiotics just before, during and 6 hours after surgery is indicated. The tendon repair is performed under axillary block anesthesia along with a pneumatic tourniquet applied at the level of the arm. The tourniquet is released Just before the wound closure in order to perform hemostasis and hence preventing formation of hematoma, infection, potential adhesion and fibrosis.

2.2 Incisions for exposure

In order to provide better visibility and to allow atraumatic repair of injured tendons, the skin wound should be debrided and enlarged. The position of the finger and the shape of the initial lesion govern the method of extension of wound for exposure; a palmar zigzag approach (Bruner's incision) can be used to extend an oblique skin laceration; a midlateral approach can be used to extend transverse skin laceration. A palmar zigzag incision provides an excellent exposure but at the same time can lead to adhesions as well as scar tissue formation over the repaired tendon. The midlateral approach allows a direct repair of the injured flexor tendon and also preserves vascular transverse branches of neurovascular bundle. Mid-palmar incisions and straight incisions that cross flexor creases should be avoided, as sharp angle in the raised skin flaps can result in tip necrosis. The wound needs to be extended in distal direction if tendon injury has occurred while fingers where in flexion and similarly wound needs to be extended proximally if at time of injury fingers where in extension.

Mid palmar transverse incision become sometimes necessary when proximal end of tendon has retracted to palmar level and if flexor tendon massage or flexion of wrist fails to deliver retracted proximal tendon end (**Figure 1**). The silicone tube is passed in retrograde direction into the palm, from superficialis chiasma up to proximal stump. The proximal end of injured tendon is attached to the silicone tube and pulled distally, delivering it back into wound. This procedure helps in avoiding traumatic injury to digital sheath and hence preventing adhesion formation (**Figure 2**) [13].

With an L-shaped incision (Lister's technique) the digital flexor tendon sheath is opened in between the annular pulleys [9]. to prevent bowstringing Annular pulleys (especially A2 and A4) should be spared and repaired if they are traumatized. The sheath should be closed using a fine suture material after completion of repair of severed flexor tendon. In case of severe damage to the sheath, it may be necessary to excise the portion of the sheath over the repaired tendon site to prevent trigger finger or an impingement. In cases where tendon injury is not a result of sharp cut, the tendon ends needs to be refreshed using a sharp blade, this debridement should be minimal to avoid any tension over the tenorrhaphy site. A needle can be used to fix proximal end in place while performing tenorrhaphy, this allows tension free approximation of two injured tendon ends (**Figure 3**).

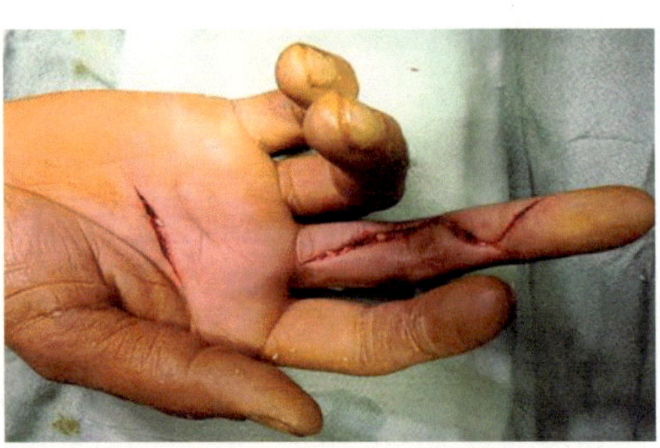

Figure 1.
Bruner's incision with palmar incision to expose the tendon stumps.

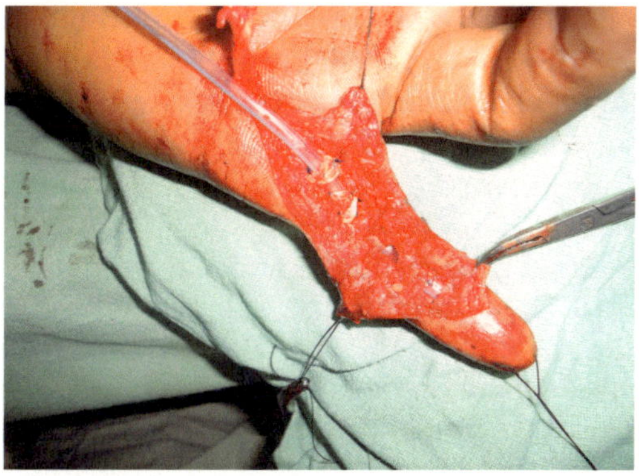

Figure 2.
A silicone tube to bring the proximal tendon stump to the desired level to avoid potential formation of adhesions.

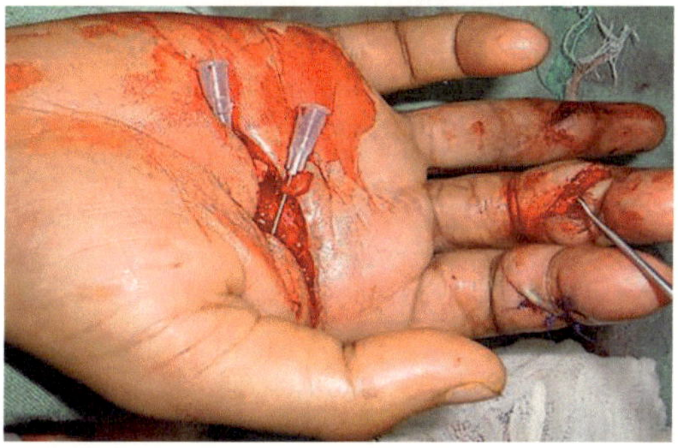

Figure 3.
Temporary fixation of the proximal tendon stump with a needle to facilitate repair.

2.3 Suturing technique

Various suturing techniques have been defined. Among these the modified Kessler suture using two sutures [14] and a 'grasping' suture [15, 16] having knots inside the cross-section have been widely accepted. A running fine epitendinous sutures increase the tensile strength and also allows smooth gliding of the repaired tendon within the digital sheath. However, immediate active rehabilitation is not possible after using these suturing techniques. Therefore, number of studies has been carried out to improve the suturing technique as well as the suturing material.

The 'ideal' suturing material should be strong, pliable but non-reactive, and of small caliber. It is advisable to use Nylon 3/0 for the central suture and for the epitendinous running suturing its recommended to use nylon 5/0 or 6/0 [17]. The 'locking' or a 'grasping' suture [18, 19] with four or six strand sutures [20, 21] is considered to be ideal for central core suturing along with running epitendinous locking sutures [22]. This allows an immediate active rehabilitation programme as this suturing technique provides double strength than usual traditional suturing methods. Tsuge's suture [23] is easy method of performing 'locking' sutures, but it leaves a knot outside over the tendon repair site, which in turn can affect smooth gliding of flexor tendons within the synovial sheath or the pulley system. This new suturing method by virtue of strong repair, allows early active motion with minimal risk of gap formation or early tendon rupture. A new material has been reported which is characterized by its high traction resistance: it consists of two intratendinous, stainless steel anchors that are joined by a multifilament stainless steel suture.

This permanent implant is intended to hold the repaired ends of tendon in close approximation until healing is completed. Protected passive tendon mobilization exercises are carried out after completion of tendon repair.

2.4 Postoperative management

postoperatively, A dorsal plaster splint is applied, from the proximal forearm to the fingertips in 'intrinsic position': the wrist is kept with 20° of palmar flexion, the MP joints in 60° flexion while as PIP and DIP joints are placed in full extension in order to avoid development of any flexion contracture. Two possible options as post-operative protocol are: to immobilize operated hand for 4 weeks or to start early mobilization according to specific exercise protocol. Immobilization is better option in non-cooperative patients and in case of children, in these patients mobilization is started in fifth week, with combination of both active as well as passive motions with dorsal blocking splint in place; seventh week onwards mobilization against resistance is initiated. The complications like tendon rupture or gaping of repair are very low with this protocol, but there are increased chances of adhesion formation, which usually requires tenolysis. Strickland and Glogovac [24] and Lister et al. [25] studied the benefits of early mobilization for tendon healing with better final end results especially in zone II flexor tendon repairs. In these studies using controlled passive mobilization post-operatively excellent results were obtained in 36% patients, 24% patients had poor results and only 4% had tenorrhaphy ruptures; while as no excellent results were obtained in cases of immobilization protocol, poor results were observed in 44% patients and 16% had tenorrhaphy site ruptures. Gelberman et al. [26, 27] also obtained superior results with early mobilization protocol and additional advantages like: improved tendon gliding (due to low rate of soft tissue adherence), enhancement of intrinsic healing mechanism, with enhanced tensile strength thereby decreasing risk of gap formation. The highest risk of tenorrhaphy rupture is between 5th and 10th postoperative day, during this period the hand therapist should be very cautious while doing during active motion exercises.

Kleinert et al. [28] proposed active and passive mobilization with dorsal blocking plaster splint keeping wrist in flexion of 20°, MP joint in flexion of 70° and allowing complete extension of fingers. An elastic traction band is attached to a loop, which is fixed to nail, keeping fingers in flexion but at same time allowing active extension within the range of dorsal blocking splint (**Figure 4**). D first 4 weeks, the patients is asked to perform active extension of the fingers many times for half an hour periods every day at different intervals. For the rest of the day and during the night the rubber band traction is detached in order to prevent development of flexion contracture in interphalangeal joint. At the beginning, the exercises should be guided by the hand therapist keeping patients elbow flexed and pronated in order to relax the flexor muscles. Between the fifth and sixth post-operative weeks, active flexion is begun with dorsal blocking splint in place. This technique is excellent but highly demanding for the therapist, surgeon as well as patient, and a control at every step is necessary to prevent a rupture or a gap at the tendon repair site. A palmar pulley situated at the level of the distal palmar crease significantly improves the range of flexion of the fingers and hence better results have been reported. After repairing the flexor pollicis longus (FPL) in the thumb MP and IP joints are kept in 20° of flexion.

Duran and Hauser [29] proposed controlled passive motion for the post-operative flexor tendon repaired lesions in zone II. The wrist is kept in 20–30° of flexion, the MP joint in 60° of flexion while as PIP and DIP joints are placed in extension.

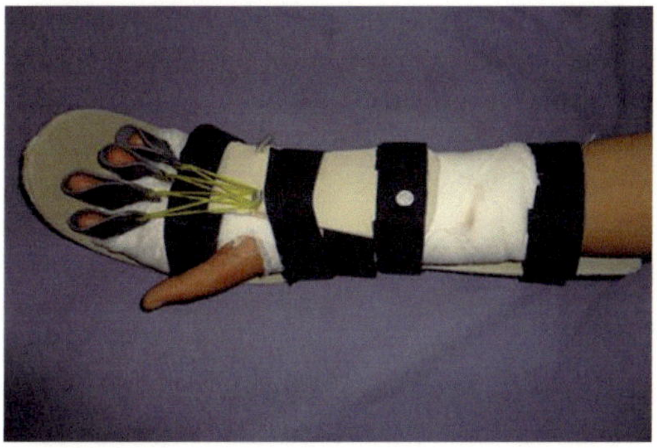

Figure 4.
Kleinert's technique for passive flexion and active extension.

For first 4 weeks controlled passive motion is used, by the hand therapist, twice a day with each session of six to eight motions for each tendon. This method uses 3–5 minute exercise movements at the repair site for preventing any firm adhesion formation. For a week, a rubber band traction is attached to the wrist and active exercises are done for 2 weeks with dorsal blocking splint in place.

Duran's technique was modified by Strickland [15]. He increased the duration and frequency of the passive daily exercises. The PIP and DIP joints are separately mobilized with repeated motions of full passive extension and flexion. An occupational therapist works closely with the operating hand surgeon and guides the controlled passive motion protocol during first 5 weeks. For starting the active flexion exercises after 5th week, the block technique advised by Bunnel [30] can be utilized: the PIP joint is actively flexed while the MP joint is blocked in extension; similarly while the DIP joint is actively flexed, the PIP joint is blocked in extension. Beyond 6 weeks, If the extension of the finger is limited, dynamic splinting may be necessary. Six months is the minimal period before considering any tenolysis and this is the time period that is required to obtain complete motion (especially in children).

Excellent results were reported by Chow et al. [31] in a multicenter study carried out for zone II flexor tendon injuries. They utilized rubber band traction with a palmar pulley at distal palmar crease level thus increasing passive flexion at the MP and PIP joints. This modification increases the differential gliding between superficialis and profundus tendon and in addition increases tendon excursion in the sheath as well. Full passive extension and flexion were performed for the first 4 weeks under the supervision of a hand therapist, in addition to the active extension exercise programme against the rubber band traction. The rubber band traction is removed for the fifth and sixth week and active and passive full flexion and extension exercises are performed. To prevent development of any contracture at the level of interphalangeal joints the supervision of both hand therapist and a hand surgeon is very important.

Many authors have reported their results using early active flexion exercises after performing flexor tendon repair in zone II [32–34]. to perform this rehabilitation programme it's important to Improve the quality as well as resistance of the suture, in order to prevent rupture or a gap of the tendon at the repair site. Indications for using early motion protocol is limited to motivated and intelligent patients having clean cut tendon injury, with a specialized hand therapist working in close collaboration with a hand surgeon.

Magnetic resonance imaging (MRI) is very useful for diagnosing many postoperative complications especially in differentiating gap from adhesion formation, especially in zone II after tenorrhaphy of FDS and FDP tendon injuries.

2.5 Zone wise features of primary tendon repair

2.5.1 Zone I

Only the FDP is involved in traumatic lesion of this zone. As the intact vincular system limits the retraction of the tendon hence the severed proximal stump is easy to find. The end to end tenorrhaphy can be done directly if the distal stump is more than 1 cm in length. In case the distal stump is less than 1 cm in length, it is recommended to reinsert the proximal tendon end into the distal phalanx. It is done as double attachment: At the base of the distal phalanx to a subperiostal flap from the palmar aspect and second attachment is distally at the mid nail plate level after passing through the distal phalanx (**Figure 5**) [35]. A4 pulley should be preserved to prevent bowstringing but the A5 pulley is frequently partially opened.

Avulsion injury of the FDP is relatively a common injury in athletes but often the diagnosis is made late. As any finger can be involved in such injury but the ring finger is most commonly involved. The diagnosis is made in an emergency set up, when an athlete is unable to actively flex the distal phalanx. A lateral radiograph should be carried out routinely, to look for a small bone at the level of the PIP or DIP joint that is actually avulsed from the base of distal phalanx. Three main factors govern the prognosis in such lesion:

1. Remaining nutritional supply to the avulsed tendon and the degree of retraction of the avulsed tendon.

2. Any diagnostic delay.

3. The size of the bony fragment avulsed.

This injury is classified into three types as per Leddy and Parker [36]:

- Type I: complete destruction of the vincular system with FDP retracted into the palm. In an emergency setup, it is advisable to carry out a reinsertion of the severed FDP, but there is significant risk of adherence formation and causing dysfunction of the intact FDS. Alternatively, conservative method of not re-inserting severed FDP tendon and excising the FDP tendon with fusion of the DIP joint can be chosen.

- Type II: FDP is retracted to the level of the PIP joint and in this type the vincular system is intact. From all the three this is commonest one. A small bone

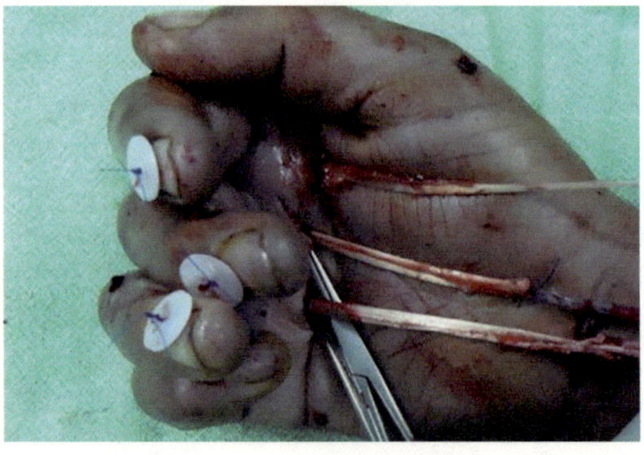

Figure 5.
Reinsertion of the avulsed tendon through the distal phalanx and fixed on the nail plate.

fragment blocked into the A3 or A2 pulley can be seen in the lateral radiograph. Early reinsertion should be performed or later but within period of 3 months provided vascular supply of the severed tendon is preserved. Satisfactory results can obtained after repair in this type.

- Type III: in this type a large bony fragment is avulsed from the distal phalanx and caught at the level of A5 or A4 pulley. As the retraction is limited hence vascularization of the severed tendon is spared. There is excellent prognosis once proper repair is done in this type of injury.

Immediately after repair an active rehabilitation programme should be started with dorsal protective splint in place.

Type IIIA lesion as reported by Robins and Dobyns [37]: In this a large bony fragment is fractured from the distal phalanx base and FDP is retracted to the level of PIP joint. Treatment comprises of reduction and an internal fixation of the distal phalanx along with reinsertion of the avulsed FDP tendon. If an avulsion injury is missed in early period, later the treatment will depend upon the degree of motivation of the patient and presence of symptoms (pain, swelling, tenderness and tumefaction) and If the patient has pain or difficulty in movement at the base of the finger it is advisable to excise the FDP tendon. Tenodesis or fusion of the DIP joint can be considered if the distal phalanx is unstable with weak pinch and an excessive dorsal extension.

Two stage flexor tendon reconstruction of the FDP tendon in zone I is indicated in selective patients; skilled technicians and musicians. It is important to explain the patients the possibility of complications like PIP joint contracture, adhesion formation or worsening of intact FDS tendon functioning.

In the thumb, if the direct repair is not possible, proximal tendon stump can be lengthened by 1–3 cm using tendinous lengthening procedures. A Z-lengthening at the wrist level [38] can give 2–3 cm of advancement and more proximally at the musculo-tendinous junction a fractional lengthening [39] produces an advancement of about 1 cm. The A2 pulley can be partially excised if required without any functional loss but Al pulley should be preserved.

2.5.2 Zone II

Traumatic lesion in this zone of the finger involves both FDP and FDS tendons, which are most often retracted into the palm (**Figures 6** and 7). It is recommended to repair both the injured tendons as repairing the FDS tendon preserves the vinculum system which ensures blood supply to the FDP tendon and in addition it maintains a smooth bed for FDP gliding. If only the FDP tendon is repaired and the FDS tendon is excised, it removes the vinculum system at the same time. However, in case both tendons are repaired there are significant chances of developing adhesions between the two tendons especially at the repair site and it may make tenolysis necessary after 6 months of repair.

In the thumb zone II injury causes the FPL trauma, which slips into the palm, and its retrieval gets difficult. In order to locate the proximal stump a small incision is made at the wrist level. This simple approach is adopted to avoid any damage to the carpal tunnel, the thenar eminence and to the cutaneous branch of the median nerve. A similar technique of passing the silicone tube as that used for retrieval of the FDS and the FDP tendons in fingers is used to bring the FPL tendon atraumatically into the initial wound for tenorrhaphy. It is recommended to preserve the A1 or oblique pulley and in case these pulleys are traumatized they need to be reconstructed using the abductor pollicis brevis aponeurosis for prevention of any bowstringing or limitation in thumb motion.

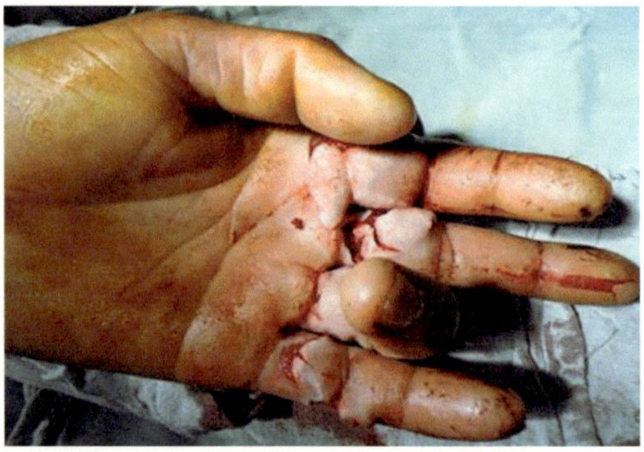

Figure 6.
Flexor tendon injuries in zone II, Bunnell's "no man's land".

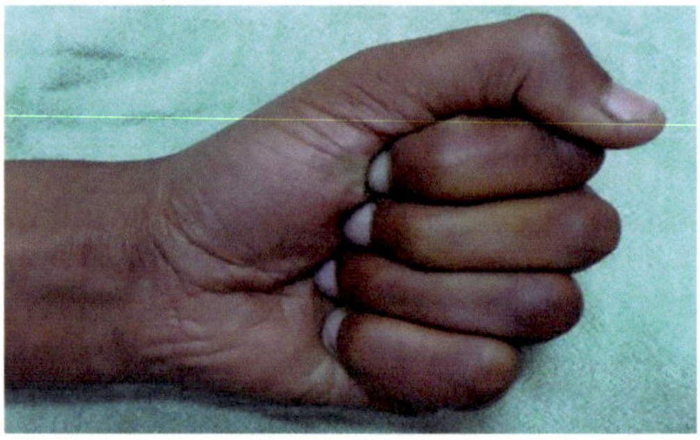

Figure 7.
Result in flexion after 6 months of modified Kessler suture and Kleinert's rehabilitation.

2.5.3 Zone III

Injury in this zone is usually associated with damage to the neurovascular bundles, which must be repaired along with the tenorrhaphy of the FDS and the FDP tendons and it is recommended to preserve the transverse palmar fascia (A0 pulley).

2.5.4 Zone IV

As this zone is protected anteriorly by the thick transverse carpal ligament hence tendon injuries are rare in this zone. But in case of crush injury in this zone, multiple tendons can get traumatized simultaneously, as they are bunched together in small space of carpal tunnel, in addition median nerve (sensory and motor branches), palmar branch of ulnar nerve can get injured as well. In order to repair all the lesions it is recommended to completely open the carpal tunnel for better access.

2.5.5 Zone V

Injury in this zone may involve ulnar artery, radial artery, median nerve, ulnar nerve and multiple tendons. The prognosis after proper repair of the injuries in zone III, IV and V is better and the return of complete motion is expected [40] after 6 months. As compared to zone II tendon adherences or tenorrhaphy rupture or need of secondary tenolysis are rare in this zone.

2.6 Difficult situations in primary tendon repair

In major trauma where flexor tendon injury is associated with phalanx fracture, skin loss, tendon loss and neurovascular lesions, regular tenorrhaphy without traction is not possible. Two options are available in case the advancement procedures are not sufficient, in zone III, IV or V a bridge graft can be used to restore tendon continuity. A regional or local flap is utilized to cover the repair site and hence decreasing the potential adhesion formation. In zone II, if pulley and the flexor tendon sheath are damaged, it makes placement of transitory silicone tube (Hunters rod) necessary for creation of new digital sheath with smooth gliding potential. This procedure is labeled as "two stage tendon reconstruction", during the first operation in addition to placement of silicone tube, repair of the neurovascular bundle and skin is done along with reconstruction of the damaged pulleys. A second operation consists of replacement of silicone rod by a tendon graft through limited exposure (Hunter procedure).

The question of whether or not to repair an isolated FDP injury in zone II, it is necessary to assess the level of retraction of proximal stump. If contusion of surrounding soft tissue is limited and the proximal stump is not retracted into the palm, a direct atraumatic repair of the FDP tendon is recommended. But if the proximal stump is retracted into palm with avulsion of vinculum system, there are high chances of local fibrosis and adhesion formation after the FDP repair, therefore in such cases it is preferable to go for DIP tenodesis or fusion rather than to primary repair of the injured FDP tendon.

In case of partial flexor tendon laceration, the surgical technique depends on the total percentage of cross-sectional area of tendon involved: if the laceration involves less than 20% of the total diameter of the tendon, the tendon should be rounded off by local flap resection; to avoid a trigger finger, a partial resection of the pulley could helpful [41]; if the laceration involves between 20 and 50% of total diameter, a running simple peripheral suture should be sufficient; if the laceration is involving more than 50% of the diameter, a core suture through the injured part of the tendon with fine peripheral running suture is recommended.

2.7 Evaluation of results after repairing flexor tendon injuries

For evaluating the results after tendon repair, many methods have been reported. American Society of Hand Surgery in 1976, proposed a method which measured the active flexion at MP, PIP and DIP joints and decreases in loss of extension for each joint. This value was then compared to the contralateral healthy finger.

Buck-Gramko [42] measured the distance from pulp-palmar, total active motion (TAM) and active extension loss of the finger. This method is lengthy and difficult to reproduce for each patient on every consultation.

Tubiana et al. [43] proposed a method of evaluation which was based on the PIP joint motion. This method evaluated the second phalanx position as compared to the metacarpal position; then the loss of active extension and active flexion of the finger can be precisely measured. This method evaluates the global function of the finger but does not measure the arc of mobility.

Strickland [16] reported a simple method, which counted the total active motion (TAM) of PIP and DIP joints. This method does not take movement across MP joint into consideration; based on the fact that flexion of MP joint is not under the control of only flexor tendons. TAM measured is compared to the contralateral finger to obtain a percentage of motion.

3. Conclusions

Many advances have been made in the understanding of tendinous healing mechanism—e.g. rehabilitation programmes, suture technique—but it is important to consider the various other factors related to injury itself that can significantly affect the final outcome, these include: the type of initial injury (clean cut, avulsion or contusion), associated injuries (phalangeal fractures, skin loss or any neurovascular bundle injury) and extent of injury to tendon sheath and pulley system, especially in zone II. However, the maneuvers which help in achieving better end results should essentially be followed, which include: an atraumatic handling of structures preferably under microscope and use of an anatomic suture with high potential of resistance thereby allowing to start immediate active motion and last but not least a well-motivated patient who understands and follows postoperative physiotherapy protocol under the supervision of a hand therapist and an operating surgeon.

Acknowledgements

We would like to thank all surgeons from department of plastic surgery at Sher-i-Kashmir Institute of Medical Sciences, Soura, Kashmir (India) for providing insight and expertise that greatly assisted in writing this chapter on "Management of Flexor Tendon Injuries in Hand".

Conflict of interest

It's to state that there was no conflict of interest in any author for writing this chapter.

Author details

Tawheed Ahmad[1*], Summaira Jan[2] and Saima Rashid[3]

1 Sher-i-Kashmir Institute of Medical Sciences, Soura, Srinagar, Kashmir, India

2 Government Dental College, Srinagar, Kashmir, India

3 Government Medical College, Srinagar, Kashmir, India

*Address all correspondence to: drtawheedahmad@yahoo.com

References

[1] Doyle JR. Anatomy of the finger flexor tendon sheath and pulley system. Journal of Hand Surgery. 1988;**13A**:473-484

[2] Doyle JR. Anatomy and function of the palmar aponeurosis pulley. Journal of Hand Surgery. 1990;**15A**:78-82

[3] Lin GT, Amadio PC, Cooney WP. Functional anatomy of the human digital flexor pulley system. Journal of Hand Surgery. 1989;**14A**:949-956

[4] Manske PR, Lesker PA. Palmar aponeurosis pulley. Journal of Hand Surgery. 1983;**8**:259-263

[5] Amiel D, Ishizue K, Billings E, et al. Hyaluronan in flexor tendon repair. Journal of Hand Surgery. 1989;**14A**:837-843

[6] Hagberg L. Exogenous hyaluronate as an adjunct in the prevention of adhesions after flexor tendon surgery. A controlled clinical trial. Journal of Hand Surgery. 1992;**17A**:132-136

[7] Gelberman RH, Woo SLY, Amiel D, Horibe SH, Lee D. Influences of flexor sheath continuity and early motion on tendon healing in dogs. Journal of Hand Surgery. 1990;**15A**:69-77

[8] Lundborg G, Hansson HA, Rank F, Rydevick B. Superficial repair of severed flexor tendons in synovial environment: an experimental, ultrastructural study on cellular mechanisms. Journal of Hand Surgery. 1980;**5**:451-461

[9] Lister G. Incision and closure of the flexor sheath during primary tendon repair. The Hand. 1983;**15**:123-135

[10] Saldana MJ, Ho PK, Lichtman DM, Chow JA, Dovelle S, Thomes LJ. Flexor tendon repair and rehabilitation in zone II: open sheath technique versus closed sheath technique. Journal of Hand Surgery. 1987;**12A**:1110-1114

[11] Peterson WW, Manske PR, Dunlap J, Horwitz DS, Kahn B. Effects of various methods of restoring flexor sheath integrity on the formation of adhesions after tendon injury. Journal of Hand Surgery. 1990;**15A**:48-56

[12] Schneider LH, Hunter JM, Norris TR, Nadeau PO. Delayed flexor tendon repair in no man's land. Journal of Hand Surgery. 1977;**2**:452-455

[13] Sourmelis SG, McGrouther D. Retrieval of the retracted flexor tendon. Journal of Hand Surgery. 1987;**12B**:109-111

[14] Tajima T. History, current status, and aspects of hand surgery in Japan. Clinical Orthopaedics. 1984;**184**:41-49

[15] Strickland JW. Results of flexor tendon surgery in zone II in flexor tendon surgery. Hand Clinics. 1985;**1**:167-169

[16] Strickland JW. Flexor tendon surgery Part I: Primary flexor tendon repair. Journal of Hand Surgery. 1989;**14R**:261-272

[17] Taras JS, Raphael JS, Marczyk SC, Bauerle WB. Evaluation of suture caliber in flexor tendon repair. Journal of Hand Surgery. 2001;**26**:1100-1104

[18] Barrie KA, Tomak SL, Cholewicki J, Merrell GA, Wolfe SW. Effect of suture locking and suture caliber on fatigue strength of flexor tendon repair. Journal of Hand Surgery. 2001;**26A**:340-346

[19] Wada A, Kubota H, Miyanishi K, et al. The mechanical properties of locking and grasping suture loop configurations in four strand core sutures technique. Journal of Hand Surgery. 2001;**25B**:548-551

[20] Savage R, Risitano G. Flexor tendon repair using a "six strand" method of repair and early active mobilisation. Journal of Hand Surgery. 1989;**14B**:396-399

[21] Dinopoulos H, Boyer MI, Burns ME, Gelberman RH, Silva MJ. The resistance of four and eight strand suture technique to gap formation during tensile testing: an experimental study of repair of canine flexor tendons after 10 days of in vivo healing. Journal of Hand Surgery. 2000;**25A**:489-498

[22] Wade PJF, Wetherell RG, Amis AA. Flexor tendon repair: significant gain in strength from the halsted peripheral suture technique. Journal of Hand Surgery. 1989;**14B**:232-235

[23] Tsuge K, Ikuta Y, Matsuhi Y. Intratendinous tendon suture in the hand; a new technique. Journal of Hand Surgery. 1975;**7**:250-255

[24] Strickland JW, Glogovac SV. Digital function following flexor tendon repair in zone II: a comparison of immobilization and controlled passive motion techniques. Journal of Hand Surgery. 1980;**6**:537-550

[25] Lister GD, Kleinert HE, Kutz JE, Atasoy E. Primary flexor tendon repair followed by immediate controlled mobilization. Journal of Hand Surgery. 1977;**2**:441-452

[26] Gelberman RH, Siegle DB, Savio LY, et al. Healing of digital flexor tendons: Importance of the interval from injury to repair. The Journal of Bone and Joint Surgery. 1991;**73A**:66-75

[27] Gelberman RH, Steinberg D, Amiel D, Akeson W. Fibroblast chemotaxis after tendon repair. Journal of Hand Surgery. 1991;**l6A**:688-693

[28] Kleinert HE, Kutz JE, Ashbell TS, Martinez E. Primary repair of lacerated flexor tendons in "No man's land". Journal of Bone and Joint Surgery. 1967;**49A**:577

[29] Duran RJ, Hauser RG. Controlled passive motions following flexor tendon repair in zones two and three. In: AAOS Symposium on Tendon Surgery in the Hand. St Louis: Mosby; 1975. pp. 105-111

[30] Bunnell S. Surgery of the Hand. 3rd ed. Philadelphia: JB Lippincott; 1956

[31] Chow A, Thomes LJ, Dovelle SW, Milnor WH, Jackson JP. Controlled motion rehabilitation after flexor tendon repair and grafting: A multi-centre study. Journal of Bone and Joint Surgery. 1988;**70B**:591-595

[32] Becker H, Orak F, Duponselle E. Early active motion following a bevelled technique of flexor tendon repair: Repair of fifty cases. Journal of Hand Surgery. 1979;**4**:454-460

[33] Small JO, Brennen MD, Colville J. Early active mobilization following flexor tendon repair in zone 2. Journal of Hand Surgery. 1989;**14B**:383-389

[34] Bellemère P, Chaise F, Friol JP, Gaisne E, Le Lardic C. Résultats de la mobilisation active précoce après reparation primaire des tendons fléchisseurs. La Main. 1998;**3**:221-233

[35] Foucher G, Merle M, Ph H. Suture du tendon fléchisseur profond au niveau de la partie distale du "No Man's land". Revue de Chirurgie Orthopédique. 1986;**72**:227-229

[36] Leddy JP, Parker JW. Avulsion of the profundus tendon insertion in athletes. Journal of Hand Surgery. 1977;**2**:66

[37] Robins PR, Dobyns JH. Avulsion of the insertion of the flexor digitorum profundus tendon associated with fracture of the distal phalanx. A brief review. In: AAOS Symposium on Tendon Injury in the Hand. St Louis: CV Mosby; 1975. pp. 151-156

[38] Urbaniak J. Repair of flexor pollicis longus. Hand Clinics. 1985;**1**:69-76

[39] Leviet D. Flexor tendon lengthening by tenotomy at the musculotendinous junction. Annals of Plastic Surgery. 1986;**17**:239-246

[40] Rogers GD, Henshall AL, Sach RP, Wallis KA. Simultaneous laceration of the median and ulnar nerves with flexor tendons at the wrist. Journal of Hand Surgery. 1990;**15A**:990-995

[41] Frewin PR, Scheker LR. Triggering secondary to an untreated partially cut flexor tendon. Journal of Hand Surgery. 1989;**14B**:419-421

[42] Buck-Gramko D. A new method for evaluation of results in flexor tendon repair. Handchirurgie. 1976;**8**:65-69

[43] Tubiana R, Gordon S, Grossman J, McMeniman P. Evaluation des résultats après réparation des tendons fléchisseurs des doigts. In: Tubiana R, editor. Traité de Chirurgie de la Main. Vol. 3. Paris: Masson; 1986. pp. 281-286

Chapter 9

The Injectable rhBMP-2-containing Collagen Gel for Tendon Healing in a Rabbit Extra-Articular Bone Tunnel Model

Kwang-Il Lee, Ju-Woong Jang and Kwang-Won Lee

Abstract

This rabbit animal study has a hypothesis that the collagen gel, which is injectable easily, can be an effective carrier for recombinant human bone morphogenetic protein 2 (rhBMP-2) for the tendon healing in a bone tunnel. The cut upper long digital extensor tendon of each rabbit was inserted into the proximal tibia bone tunnel, and rhBMP-2 conjugated collagen gel was injected into the tendon-bone tunnel interface using a syringe. Biomechanical and histological performances were analyzed at 3 and 6 weeks after surgery. The collagen sol at room temperature was transformed to a gel at 37°C. The rhBMP-2 was slowly released from the collagen gel for more than 4 weeks. The in vivo experiment showed the enhanced new fibrocartilage and bone tissue formation at 6 weeks after injecting the rhBMP-2-containing collagen gel. The calcification and enthesis-like tissue were detected radiologically in the repaired tendon-bone junction. The viscous collagen gel-containing rhBMP-2 increased the fusion rate of the repaired tendon in the bone tunnel. This study showed that viscous collagen gel can be an effective carrier for rhBMP-2 for tendon healing in the bone tunnel. The rhBMP-2-containing collagen gel will be promising for tendon-bone interface healing in the future.

Keywords: rhBMP-2, tendon, bone tunnel, enthesis, ligament injury

1. Introduction

Tendon healing in the bone tunnel is the one of the most critical points for tendon repair [1]. Various types of tendon grafts such as peroneus longus, tibialis, gracilis, semitendinosus, and Achilles tendons can be transplanted for replacing the ruptured ligament/tendon tissues [2]. However, the failure rate after surgery has still remained the cause of poor recovery capacity, so more strategic studies are necessary [3]. One of the main reasons for poor healing is because the mechanical stresses keep affecting repaired tendon-bone tunnel junction [4]. The unique transitional tissue, the enthesis which is a fibrocartilage tissue, is generated with a connected region between tendon and bone tissue [5]. Thus, for successful tendon/ligament reconstruction, osteointegration of inserted tendon grafts in the bone tunnel is strongly recommended [6].

Recombinant human bone morphogenetic protein-2 (rhBMP-2) is a well-known growth factor for new bone formation [7–9]. This promotes the differentiation of undifferentiated mesenchymal cells into chondrogenic or osteogenic lineages that support new bone formation [10, 11]. This differentiation also needs in the reconstructed tendon-bone tunnel region for postnatal enthesis formation. There have been many studies about an effective rhBMP-2 delivery system using collagen-based materials such as sponge and gel for sustained release [12–15]. In the previous studies, rhBMP-2 was used for bone-tendon interface healing using a collagen sponge [16, 17]. However, the collagen sponge system had a limitation of rhBMP-2 localization to the only targeted region and minimizing leakage from the bone tunnel [18].For the enhanced delivery of rhBMP-2 into the targeted surgical sites, a viscous collagen gel may be useful. Another previous study had performed a comparative study of the osteogenic effects using two different rhBMP-2 delivery systems such as collagen sponge and collagen gel in a rat spinal fusion model [19]. The results showed that rhBMP-2 containing a collagen sponge had side effects of leakage BMP. To overcome this limitation, viscous collagen gel was applied to rabbit tendon-bone tunnel regions in this study.

The purpose of this translational study is to investigate whether the rhBMP-2-containing collagen gel can localize in the surgical site and improve new enthesis formation within reconstructed tendon-bone tunnel after the surgery.

2. Materials and methods

2.1 Conjugation of collagen gel and rhBMP-2

1% (w/v) collagen gel originated from porcine skin was mixed with 50 μg/ml rhBMP-2. This concentration had been confirmed from our previous rat and rabbit animal studies [20–23]. To check temperature dependency of collagen's sol-gel phase transition, the optical density of 1% collagen gel at 37°C was read at 313 nm in an absorbance microplate reader at 10, 20, and 30-minute time points. For the release kinetics analysis of rhBMP-2-containing collagen gel, it was plated on 12-well plate and incubated in 1 ml phosphate-buffered saline (pH 7.4) at 37°C for 28 days. At each time point of 1, 3, 5, 7, 14, and 28-day time points, each supernatant was collected and stored at −80°C until reading. Then, the rhBMP-2 was quantitated using an enzyme-linked immunosorbent assay kit and a cumulative release curve was plotted.

2.2 Animal study design and operative procedure

Healthy adult New Zealand White rabbits (n = 36, 3.0–3.5 kg) were used for this study. The animal treatment was followed by the Guidelines for Care and Use of Laboratory Animals, and this animal experiment was approved by the Committee of Experimental Animal Sciences. The rabbits were classified with three different groups: saline injection only (control group), collagen gel injection only without rhBMP-2 (collagen gel group), and rhBMP-2-conjugated collagen gel injection (rhBMP-2-collagen gel group). Rabbits were anesthetized with ketamine, 40 mg/kg IM; xylazine, 5 mg/kg IM.

The rabbits underwent an operative procedure for an extra-articular tendon-bone healing model at the rerouted long digital extensor tendon. The knee joint was accessed through a lateral para-patellar incision. The long digital extensor tendon was identified and then detached from its insertion at the lateral femoral condyle by sharp dissection. The free tendon was tied with 3-0 Vicryl. Then, the anterior tibia muscle was retracted laterally. A bone tunnel was created in the proximal tibia metaphysis with 30° angle relative to the long-bone axis using a drill (diameter: 2 mm). The finalized bone tunnel size after drilling was an average of 2.09 ± 0.04 mm-diameter and 5.13 ± 0.05 mm-length, which was measured by scanned microcomputed tomography images. The cut long digital extensor tendon was relocated. It was pulled manually through the bone tunnel and sutured to the periosteum and soft tissue at the medial proximal tibia with 3-0 nylon (**Figure 1**).

Each 200μl of rhBMP-2-containing collagen gel was injected into the tendon-bone tunnel junction. The joint capsule, fascia, and subcutaneous tissue were closed with 3-0 Vicryl and skin was closed with 3-0 nylon.

2.3 Analysis of three-dimensional computed tomography (CT) and bone mineral density (BMD)

The BMD and mineralized tissue ingrowth inside the tendon-bone tunnel junction were quantified by using CT system. Specimens were scanned perpendicular to the long-bone axis covering the entry and exit of the bone tunnel. The sections were reconstructed using the 3D software. To quantify the amount of newly formed mineralized tissue over time, the regions of interest (ROI) was chosen and

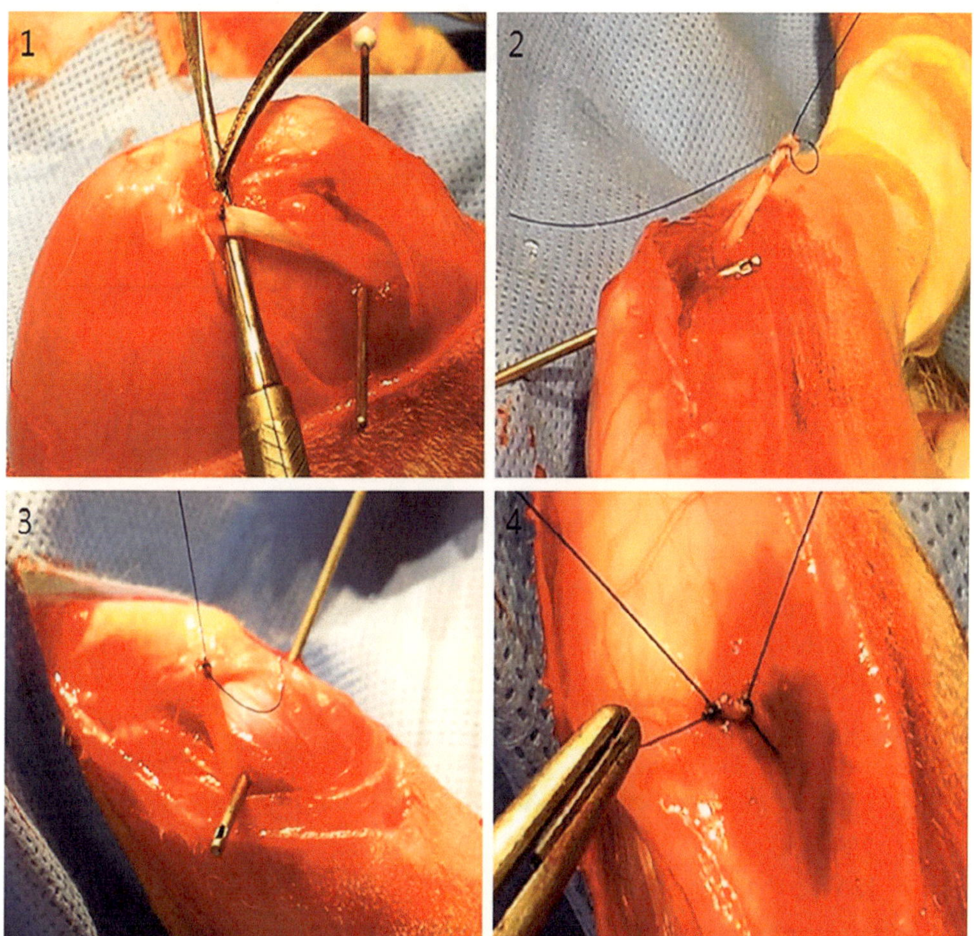

Figure 1.
Operative procedure of the long digital extensor tendon sutured to the periosteum and soft tissue of rabbit medial tibia.

reconstructed using the 3D software. After thresholding, the BMD (mg/cm^3) of the mineralized tissue inside the tendon-bone tunnel junction was calculated.

2.4 Biomechanical testing

Rabbit knee joints including the tendon-bone tunnel site were collected. To analyze the tensile mechanical properties, tensile strength was measured by a universal testing machine. The specimen was fixed vertically on a 5000 N load-cell, and tensile strength was measured by pulling each specimen at a load displacement rate of 10 mm/min. The failure load and ultimate strength (N) were recorded.

2.5 Histologic and histomorphometric analyses

Rabbit knee joints were collected and fixed in a neutralized formalin solution for 2 days and decalcified in 10% formic acid until being cut. The specimens were sliced into 4-um-thickness in an orientation parallel to the bone tunnel, and each section was stained with Masson's trichrome, and visualized using an optical microscope. Healing of the tendon in the bone tunnel was graded histomorphologically by two blinded observation methods.

Histomorphometric analysis was assessed for tendon healing in the bone tunnel. Quantitative histomorphometric analysis was performed by two blinded observations, which apportioned 0–3 points based on histomorphologic criteria, representing fibrocartilage formation, new bone formation, and tendon graft bonding to adjacent tissue (**Table 1**).

Characteristic	Points
Fibrocartilage formation	
Abundant	3
Moderate	2
Slight	1
None	0
New bone formation	
Abundant	3
Moderate	2
Slight	1
None	0
Tendon graft bonding to adjacent tissue	
75–100%	3
50–75%	2
25–50%	1
0–25%	0

Table 1.
Histomorphometric analysis to assess healing of the tendon within the bone tunnel (full score = 9 points).

2.6 Statistical analysis

The data were averaged from at least triplicated samples. The same experiments were repeated for three times to ensure the reproducibility of the methods used. All statistical analyses were performed using SPSS version 15.0. The post hoc Scheffé test was sued to analyze the significant difference between groups, with significance levels at *p < 0.05 and ** < 0.01.

3. Results

The turbidity of collagen gel at 37°C was significantly increased between the time points of 10 (OD: 0.953) and 20 (OD: 4.099) minutes (**Figure 2**). The half release from the total rhBMP-2 quantity from the collagen gel was done within 5 days from incubation. And the rest half was released slowly for over 28 days when it was released 89.3% totally. The rhBMP-2 conjugated collagen gel showed a sustained release phase (**Figure 3**).

In the 3D-CT analysis, new bone formation was detected at the interface between the tendon and tibia bone tunnel in the rhBMP-2-collagen gel group after 3 weeks even though the normal distal epiphyseal plate of rabbits has a limited cancellous bone. However, the control and collagen gel groups did not show new bone formation (**Figure 4**). At 6 weeks after the surgery, the rhBMP-2-collagen gel group

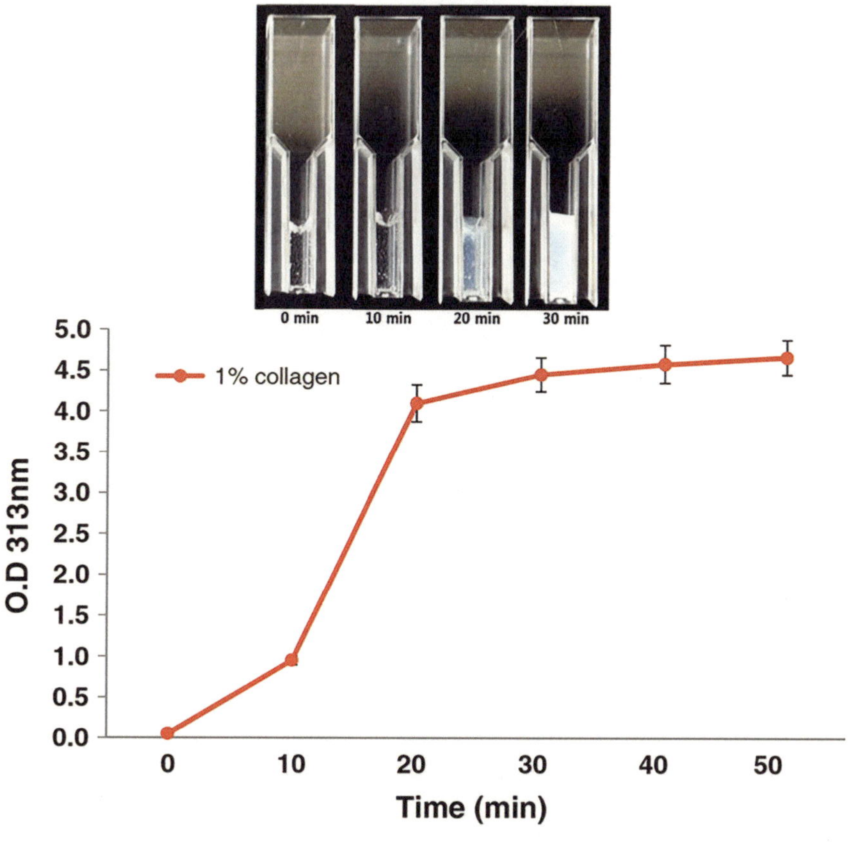

Figure 2.
Ultraviolet turbidity of 1% collagen gel at each time point, dependent on temperature; average of triplicates at each time point, which is 0, 10, 20, and 30 min at 37°C incubation.

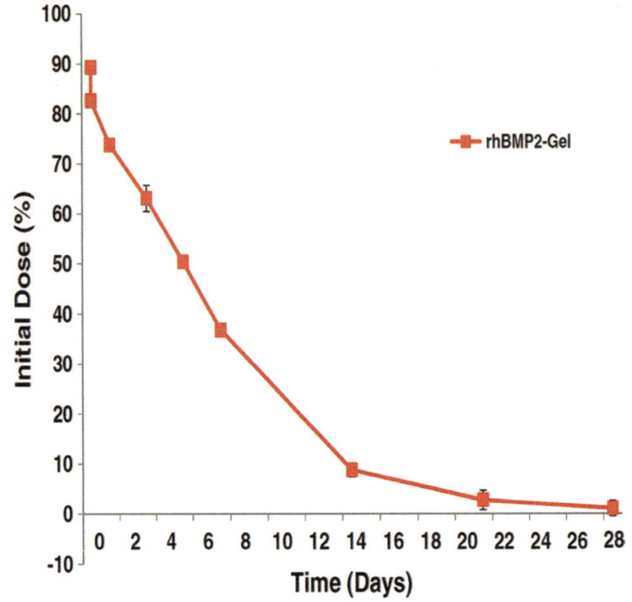

Figure 3.
Release profile of rhBMP-2 from 1% collagen solution; average of triplicates at each time point for 4 weeks.

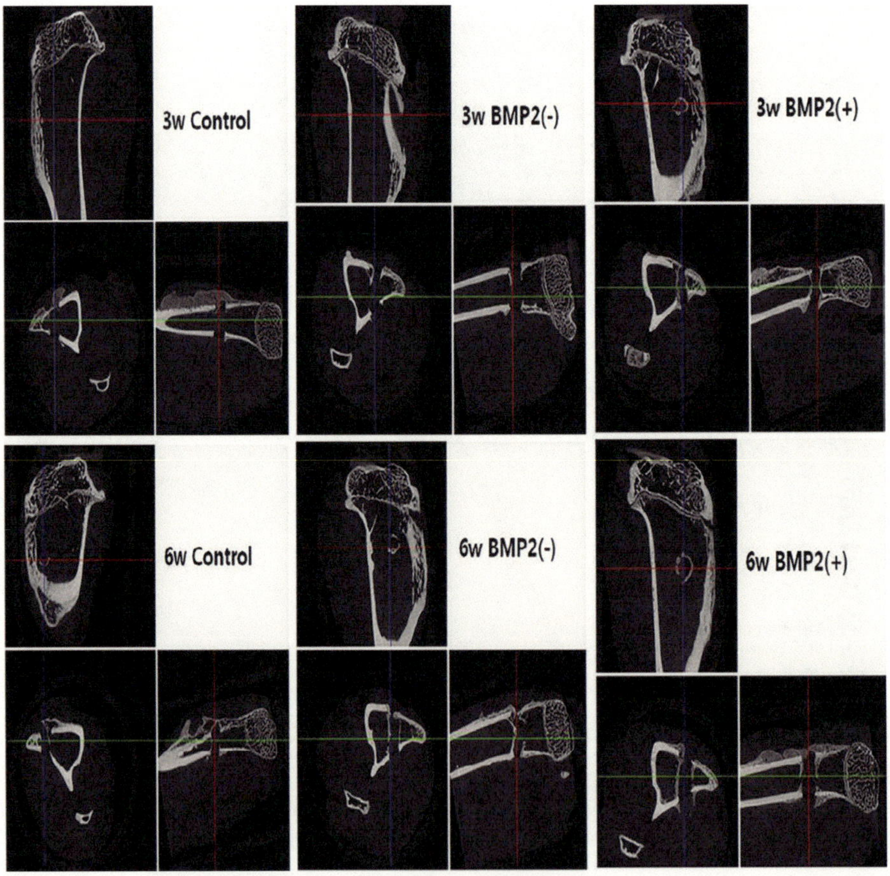

Figure 4.
3D CT images of the enthesis generated by transfer of the toe flexor or rhBMP-2$^+$ or rhBMP-2$^-$ bone complex to the proximal tibia at 3 and 6 weeks.

showed higher new bone formation than the other groups. In addition, the BMD of the rhBMP-2-collagen gel groups was significantly higher than the control group at 3 and 6 weeks after the surgery (**Figure 5**). The BMD of the collagen gel group was slightly higher than the control group; however, it was not significant.

In the biomechanical test result, the ultimate failure load of the rhBMP-2-collagen gel group was significantly higher than the other groups at 3 and 6 weeks (**Figure 6**). After 3 weeks, the ultimate failure load of the rhBMP-2-collagen gel group was 2.5-fold higher than the control group. After 6 weeks, the thBMP-2-collagen gel group was 1.8-fold higher than the control group. However, there was no significance between the collagen gel group and control group.

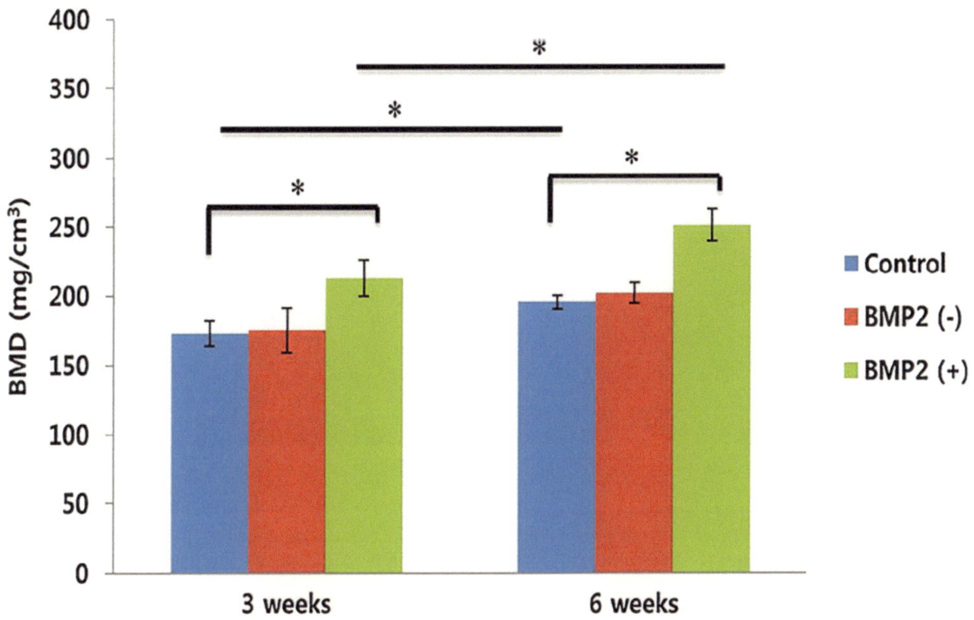

Figure 5.
Bone mineral density of the enthesis generated by transfer of the toe flexor or rhBMP-2^{+} or rhBMP-2^{-} bone complex to the proximal tibia at 3 and 6 weeks.

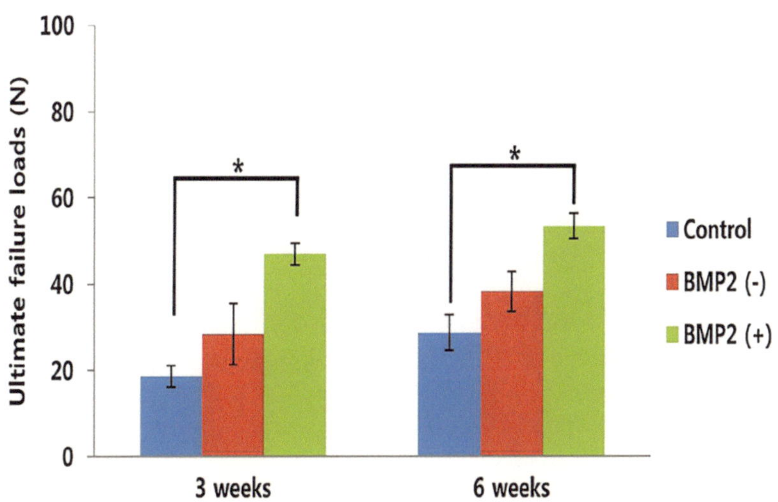

Figure 6.
Ultimate failure loads of the enthesis generated by transfer of the toe flexor or rhBMP-2^{+} or rhBMP-2^{-} bone complex to the proximal tibia at 3 and 6 weeks.

In the histology results, Masson's trichrome staining showed that collagen fibers and fibrous cartilage were widely detected in the tendon-bone interface of the rhBMP-2-collagen gel group at 3 weeks. The new bone was partly detected between the tendon and host bone (**Figure 7**). After 6 weeks, there was increased fibrous cartilage and new bone tissues between the tendon and the bone. Moreover, the new Sharpey-like fibers were detected in the rhBMP-2-collagen gel group (**Figure 8**).

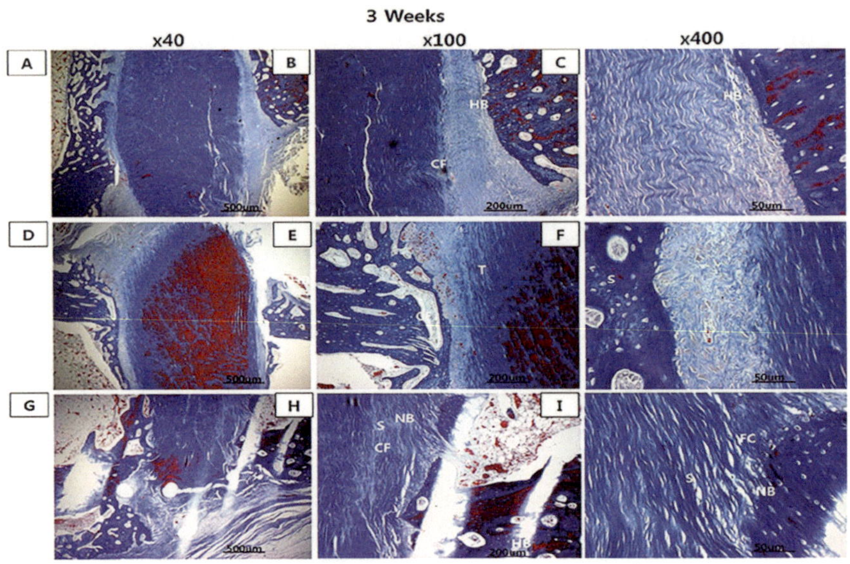

Figure 7.
Masson's trichrome staining of the enthesis generated by transfer of the toe flexor or rhBMP-2$^+$ or rhBMP-2$^-$ bone complex to the proximal tibia at 3 weeks: (A–C) control; (D–F) collagen only; (G–I) collagen with rhBMP-2; CF, collagen fibers; FC, fibrocartilage; HB, host bone; NB, new bone; S, Sharpey-like fibers; and T, tendon.

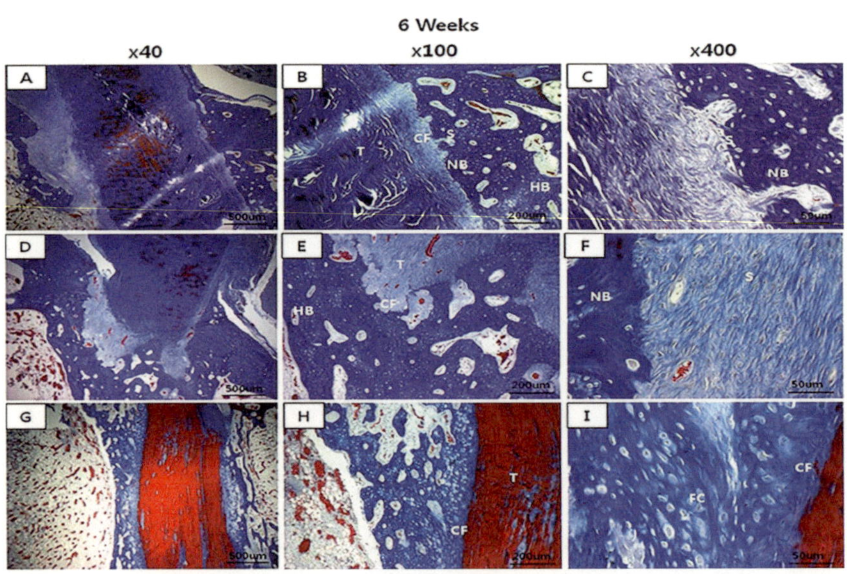

Figure 8.
Masson's trichrome staining of the enthesis generated by transfer of the toe flexor or rhBMP-2$^+$ or rhBMP-2$^-$ bone complex to the proximal tibia at 6 weeks: (A–C) control; (D–F) collagen only; (G–I) collagen with rhBMP-2; CF, collagen fibers; FC, fibrocartilage; HB, host bone; NB, new bone; S, Sharpey-like fibers; and T, tendon.

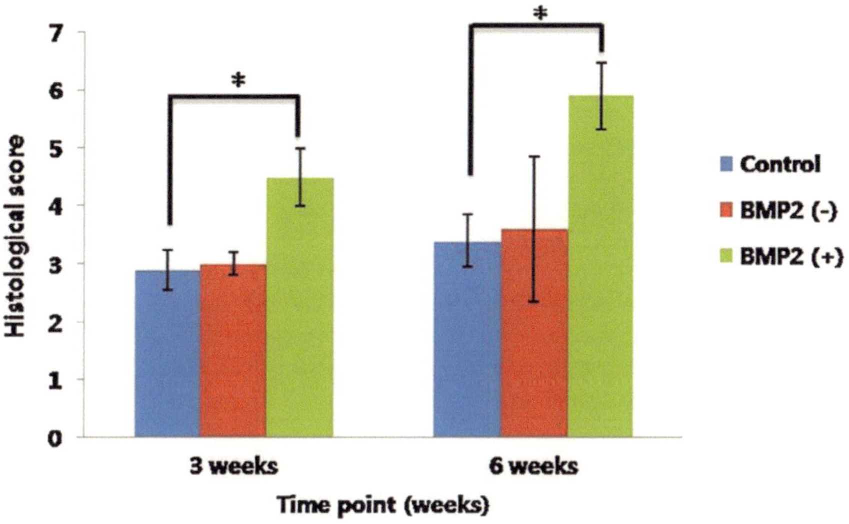

Figure 9.
Histological score of the enthesis generated by transfer of the toe flexor or rhBMP-2$^+$ or rhBMP-2$^-$ bone complex to the proximal tibia at 3 and 6 weeks.

In the histomorphometric analysis results, the histologic score about enthesis formation in the rhBMP-2-collagen gel group was significantly higher than the other groups (**Figure 9**). Moreover, the results after 6 weeks were higher in score in the rhBMP-2-collagen gel group than the score in the other groups after 3 weeks.

4. Discussion

The healing process after tendon or ligament reconstruction needs stable enthesis generation at the interface between the inserted tendon and drilled bone tunnel, which is one of the most important conditions [4, 24]. rhBMP-2 may be useful as a growth factor for new bone formation by inducing differentiation of osteoprogenitor cells to osteoblasts [25]. For the effective soft tissue healing, rhBMP-2 application will be the ideal method for new bone formation between the inserted tendon and the drilled bone tunnel. However, rhBMP-2 requires a carrier for embedding itself [26, 27]. It is important to establish the rhBMP-2 delivery system with immobilization for the sustained release in the surgical site. The immobilization of rhBMP-2 can enhance the host cell infiltration into the surgical site and stimulating cellular activity for new tissue formation [28, 29].

We used viscous and elastic collagen gel for the minimum loss and sustained release of rhBMP-2. Collagen gel is easy to inject and biocompatible for drug delivery. Collagen gel also has been employed as a scaffold in tissue engineering and regenerative medicine [30]. When the collagen gel is injected into the surgical site, it will be easy to use due to its viscous solution state. After its implantation, the gel becomes semisolid at body temperature. This thermos-sensitive state can make the injected gel stable and good for sustained release of a growth factor for soft tissue reconstruction.

In this study, a rabbit extra-articular bone tunnel model was used to investigate the tendon/ligament healing in the drilled bone tunnel. We developed an advanced viscous rhBMP-2-conjugated collagen gel for the soft tissue reconstruction. This gel system will be useful as a void filler between the tendon graft and host bone tunnel.

Collagen showed temperature-responsive gelation at the body temperature. This demonstrates that collagen gel can be effective for the stable filling into the surgical site and rhBMP-2 will be effectively and slowly released from the stable gel without any loss by irrigation during the surgery. The phase transformation will also affect the degradation rate of the collagen gel and the time course of stimulation of osteogenesis [18].

In vivo test results showed that the rhBMP-2-collagen gel group increased the fusion rate between the grafted tendon and host bon tunnel. BMD analysis results also showed the enhanced new bone formation by rhBMP-2.

In conclusion, the injectable rhBMP-2-containing collage gel induced earlier and more new bone formation at the tendon-bone tunnel interface. This study demonstrated that the mixture of the rhBMP-2 and collagen gel can accelerate the healing process of the grafted tendon in the host bone tunnel. The clinical use of the injectable rhBMP-2-collagen gel will be promising for the enhancement of tendon/ligament reconstruction in the future.

Acknowledgements

Authors would like to acknowledge the support by Ministry of Trade, industry, and Energy (grant no: 10037842).

Author details

Kwang-Il Lee[1,2]*, Ju-Woong Jang[2] and Kwang-Won Lee[3]

1 The Scripps Research Institute, La Jolla, CA, United States

2 The Institute of Biomaterials and Medical Engineering, Cellumed Co., Ltd., Seoul, Republic of Korea

3 Department of Orthopedic Surgery, Eulji University Hospital, Daejeon, Republic of Korea

*Address all correspondence to: jasonklee77@gmail.com

References

[1] Chen CH. Graft healing in anterior cruciate ligament reconstruction. Sports Medicine Arthroscopy Rehabilitation Therapy & Technology. 2009;**1**(1):21. DOI: 10.1186/1758-2555-1-21

[2] Mao H, Xu G. Soft tissue repair for tibialis anterior tendon ruptures using plate and screw fixation technique in combination with anterolateral thigh flaps transplantation. Journal of Orthopaedic Surgery and Research. 2015;**10**:143. DOI: 10.1186/s13018-015-0278-5

[3] Kuo CK, Marturano JE, Tuan RS. Novel strategies in tendon and ligament tissue engineering: Advanced biomaterials and regeneration motifs. Sports Medicine Arthroscopy Rehabilitation Therapy & Technology. 2010;**2**:20. DOI: 10.1186/1758-2555-2-20

[4] Lui P, Zhang P, Chan K, Qin L. Biology and augmentation of tendon-bone insertion repair. Journal of Orthopaedic Surgery and Research. 2010;**5**:59. DOI: 10.1186/1749-799X-5-59

[5] Thomopoulos S, Genin GM, Galatz LM. The development and morphogenesis of the tendon-to-bone insertion—What development can teach us about healing. Journal of Musculoskeletal & Neuronal Interactions. 2010;**10**(1):35-45

[6] Bi F, Shi Z, Liu A, Guo P, Yan S. Anterior cruciate ligament reconstruction in a rabbit model using silk-collagen scaffold and comparison with autograft. PLoS One. 2015;**10**(5):e0125900. DOI: 10.1371/jounal.pone.0125900

[7] Wozney JM. Overview of bone morphogenetic proteins. Spine (Philla Pa 1976). 2002;**27**(16 Suppl 1):S2-S8

[8] Riley EH, Lane JM, Urist MR, Lyons KM, Lieberman JR. Bone morphogenetic protein-2: Biology and applications. Clinical Orthopaedics and Related Research. 1996;**324**:39-46

[9] Cochran DL, Schenk R, Buser D, Wozney JM, Jones AA. Recombinant human bone morphogenetic protein-2 stimulation of bone formation around endosseous dental implants. Journal of Periodontology. 1999;**70**(2):139-150

[10] Schmitt B, Ringe J, Häupl T, Motter M, Manz R, Burmester GR, et al. BMP2 initiates chondrogenic lineage development of adult human mesenchymal stem cells in high-density culture. Differentiation. 2003;**71**(9-10):567-577

[11] Hassan MQ, Tare RS, Lee SH, Mandeville M, Morasso MI, Javed A, et al. BMP2 commitment to the osteogenic lineasge involves activation of Runx2 by DLX3 and a homeodomain transcriptional network. The Journal of Biological Chemistry. 2006;**281**(52):40515-40526

[12] Lo KW, Ulery BD, Ashe KM, Laurencin CT. Studies of bone morphogenetic protein-based surgical repair. Advanced Drug Delivery Reviews. 2012;**64**(12):1277-1291. DOI: 10.1016/j.addr.2012.03.014

[13] Boerckel JD, Kolambkar YM, Dupont KM, Uhrig BA, Phelps EA, Stevens HY, et al. Effects of protein dose and delivery system on BMP-mediated bone regeneration. Biomaterials. 2011;**32**(22):5241-5251. DOI: 10.1016/j.biomaterials.2011.03.063

[14] Lee CH, Singla A, Lee Y. Biomedical applications of collagen. International Journal of Pharmaceutics. 2001;**221**(1-2):1-22

[15] Geiger M, Li RH, Friess W. Collagen sponges for bone regeneration with

rhBMP-2. Advanced Drug Delivery Reviews. 2003;**55**(12):1613-1629

[16] Rodeo SA, Suzuki K, Deng XH, Wozney J, Warren RF. Use of recombinant human bone morphogenetic protein-2 to enhance tendon healing in a bone tunnel. The American Journal of Sports Medicine. 1999;**27**(4):476-488

[17] Thomopoulos S, Kim HM, Silva MJ, Ntouvali E, Manning CN, Potter R, et al. Effect of bone morphogenetic protein 2 on tendon-to-bone healing in a canine flexor tendon model. Journal of Orthopaedic Research. 2012;**30**(11):1702-1709. DOI: 10.1002/jor.22151

[18] Patel VV, Zhao L, Wong P, Pradhan BB, Bae HW, Kanim L, et al. An in vitro and in vivo analysis of fibrin glue use to control bone morphogenetic protein diffusion and bone morphogenetic protein stimulated bone growth. The Spine Journal. 2006;**6**(4):397-403

[19] Moon KH, Park IS, Lee MO, Park HC, Park CO, Hyun DK, et al. A comparative study of the osteogenic potentials of rhBMP-2 in collagen sponge or mixed with porcine collagen gel in a rat spinal fusion model. Tissue Engineering and Regenerative Medicine. 2009;**6**(4-11):653-658

[20] Jang JW, Yun JH, Lee KI, Jang JW, Jung UW, Kim CS, et al. Osteoinductive activity of biphasic calcium phosphate with different rhBMP-2 doses in rats. Oral Surgery, Oral Medicine, Oral Pathology, Oral Radiology. 2012;**113**(4):480-487. DOI: 10.1016/j.tripleo.2011.04.013

[21] Kim JW, Jung IH, Lee KI, Jung UW, Kim CS, Choi SH, et al. Volumetric bone regenerative efficacy of biphasic calcium phosphate collagen composite block loaded with rhBMP-2 in vertical bone augmentation model of a rabbit calvarium. Journal of Biomedical Materials Research. Part A. 2012;**100**(12):3304-3313. DOI: 10.1002/jbm.a.34278

[22] Lee KW, Lee JS, Jang JW, Shim YB, Lee KI. Tendon-bone interface healing using an injectable rhBMP-2-containing collagen gel in a rabbit extra-articular bone tunnel model. Journal of Tissue Engineering and Regenerative Medicine. 2017;**11**(5):1435-1441. DOI: 10.1002/term.2041

[23] Lee KW, Lee JS, Kim YS, Shim YB, Jang JW, Lee KI. Effective healing of chronic rotaor cuff injury using recombinant bone morphogenetic protein-2 coated dermal patch in vivo. Journal of Biomedical Materials Research. Part B, Applied Biomaterials. 2017;**105**(7):1840-1846. DOI: 10.1002/jbm.b.33716

[24] Zelzer E, Blitz E, Killian ML, Thomopoulos S. Tendon-to-bone attachment: From development to maturity. Birth Defects Research. Part C, Embryo Today. 2014;**102**(1):101-112. DOI: 10.1002/bdrc.21056

[25] Takada T, Katagiri T, Ifuku M, Morimura N, Kobayashi M, hasegawa K, et al. Sulfated polysaccharides enhance the biological activities of bone morphogenetic proteins. The Journal of Biological Chemistry. 2003;**278**(44):43229-43235

[26] Tsujigiwa H, Nagatsuka H, Gunduz M, Rodriguez A, Rivera RS, Legeros RZ, et al. Effects of immobilized recombinant human bone morphogenetic protein-2/succinylated type I atelocollagen on cellular activity of ST2 cells. Journal of Biomedical Materials Research. Part A. 2005;**75**(1):210-215

[27] Luca L, Rougemont AL, Walpoth BH, Gurny R, Jordan O. The effects of carrier nature and pH on rhBMP-2-induced ectopic bone formation. Journal of Controlled Release.

2010;**147**(1):38-44. DOI: 10.1016/j.jconrel.2010.06.011

[28] Mooney DJ, Boontheekul T, Chen R, Leach K. Actively regulating bioengineered tissue and organ formation. Orthodontics & Craniofacial Research. 2005;**8**(3):141-144

[29] Li RH, Wozney JM. Delivering on the promise of bone morphogenetic proteins. Trends in Biotechnology. 2001;**19**(7):255-265

[30] Wallace DG, Rosenblatt J. Collagen gel systems for sustained delivery and tissue engineering. Advanced Drug Delivery Reviews. 2003;**55**(12):1631-1649